The interaction between light and electrons in semiconductors forms the basis for many interesting and practically significant properties. This book examines the fundamental physics underlying this rich complexity of photoelectronic properties of semiconductors, and will familiarise the reader with the relatively simple models that are useful in describing these fundamentals. The basic physics is also illustrated with typical recent examples of experimental data and observations.

Following introductory material on the basic concepts, the book moves on to consider a wide range of phenomena, including photoconductivity, recombination effects, photoelectronic methods of defect analysis, photoeffects at grain boundaries, amorphous semiconductors, photovoltaic effects and photoeffects in quantum wells and superlattices. The author is Professor of Materials Science and Electrical Engineering at Stanford University, and has taught this material for many years. He is an experienced author, his earlier books having found wide acceptance and use. This book represents, as much as any one book can, his 44 years of research to date.

Readers will therefore find this volume to be an up-to-date and concise summary of the major concepts, models and results. It is intended as a text for graduate students, but will be an important resource for anyone researching in this interesting field.

Photoelectronic properties of semiconductors

# Photoelectronic
# properties
# of semiconductors

**RICHARD H. BUBE**

*Professor of Materials Science and Electrical Engineering*
*Stanford University*

**CAMBRIDGE**
UNIVERSITY PRESS

Published by the Press Syndicate of the University of Cambridge
The Pitt Building, Trumpington Street, Cambridge CB2 1RP
40 West 20th Street, New York, NY 10011-4211, USA
10 Stamford Road, Oakleigh, Victoria 3166, Australia

First published 1992

*A catalogue record of this book is available from the British Library*

*Library of Congress cataloguing in publication data*

Bube, Richard H., 1927–
Photoelectronic properties of semiconductors/Richard H. Bube.
    p.    cm.
Includes bibliographical references.
ISBN 0-521-40491-6.—ISBN 0-521-40681-1 (pbk.)
1. Semiconductors. 2. Photoelectronics. I. Title.
TK7871.85.B83   1992
621.381'52—dc20      91-20942   CIP


ISBN 0 521 40491 6 hardback
ISBN 0 521 40681 1 paperback


Transferred to digital printing 2004

UY

*Ever since the creation of the world*
*His invisible nature, namely, His eternal*
  *power and deity,*
*has been clearly perceived*
*in the things that have been made.*

*Romans 1:20–(RSV)*

# Contents

# Preface

My earlier book, *Photoconductivity of solids*, published in 1960 and subsequently in print for 26 years, attempted to describe all of the previous work related to photoconductivity or any of its associated phenomena. From the response that it evoked it appears that this attempt was reasonably successful.

But the field continued to expand rapidly. If a thousand references were adequate to describe almost a century from Willoughby Smith's discovery of photoconductivity in selenium in 1873, tens of thousands of references have been added in recent years. A representative bibliography of books and reviews is included at the end of the text. In 1976 when *Photoconductivity and related phenomena*, edited by J. Mort and D. Pai was published, fifteen authors contributed to cover a wide variety of solids in addition to crystalline semiconductors, like molecular crystals, amorphous materials, polymeric photoconductors, and non-polar liquids, which had been mentioned either not at all or very little in *Photoconductivity of solids*. It is no longer possible for a single, reasonably long monograph to provide a complete discussion of the whole field.

So it is not the purpose of this book even to attempt to give an exhaustive treatment of all photoelectronic phenomena in semiconductors. Rather it is the purpose to provide a look at the fundamentals that underlie the rich complexity of phenomena, to provide the reader with some familiarity for relatively simple models that are useful in describing these fundamentals, and to give typical recent examples of experimental data and observations that illustrate them. The present book deliberately chooses to limit itself to photoelectronic properties of semiconductors, still a very large and rapidly growing field indeed, especially when one realizes the potential of low-dimensional structures such as quantum wells and superlattice structures.

It has been a special thrill for me to be able to write this book, in a kind of way reliving and at the same time updating my professional research career. My purpose in writing the book is facilitated by the fact that for the past 30 years I have been teaching courses related to this subject area at Stanford University. I would like to express my gratitude toward all those friends and colleagues through the years who have contributed to my progress, and to all of my students (Ph.D. No. 50 graduated in 1990) involved in research with me, as well as to those many more who interacted in the courses taught at Stanford and thereby directly or indirectly contributed to the material of this book.

I was in the midst of the task of writing the book, when I learned of the death of Dr Albert Rose, who was my professional mentor and friend particularly during the 14 years I spent at the RCA Laboratories. I wish to acknowledge my special indebtedness to him.

My beloved wife Betty was by my side through the writing of the original *Photoconductivity of solids,* and I thank God that I have had her unfailing love and support through all the intervening years.

**Richard H. Bube**
*Stanford University*
*Stanford, California*

# 1

# Introductory concepts

Many photoelectric phenomena are describable in terms of a
set of basic concepts involving electron activity in semiconductors.
These include optical absorption by which free carriers are created,
electrical transport by which free carriers contribute to the electrical
conductivity of the material, and the capture of free carriers leading
either to recombination or trapping.

## 1.1 Overview

These effects are illustrated in Figure 1.1. Intrinsic optical
absorption in Figure 1.1($a$) corresponds to the raising of an electron
from the valence band to the conduction band. Extrinsic optical
absorption corresponds to the raising of an electron from an imperfection to the conduction band as in Figure 1.1($b$), or the raising of an
electron from the valence band to an imperfection in Figure 1.1($c$).

Optical absorption is described quantitatively through the absorption constant $\alpha$. In the simplest case, neglecting reflection or interference effects, if light of intensity $I_0$ is incident on a material of
thickness $d$ with absorption constant $\alpha$, the intensity of the transmitted
light $I$ is given approximately by Beer's Law,

$$I = I_0 \exp(-\alpha d) \tag{1.1}$$

A free electron may be captured at an imperfection as in Figure
1.1($d$), or a free hole may be captured at an imperfection as in Figure
1.1($e$). The capture process is described through a capture coefficient $\beta$
such that the rate of capture $R$ of a species with density $n$ by a species
with density $N$ is given by

$$R = \beta n N \tag{1.2}$$

The capture coefficient $\beta$ is often expressed as the product of a capture cross section $S$ and the average thermal velocity $v$ of the free carrier.

$$\beta = \langle S(E)v(E) \rangle = Sv \qquad (1.3)$$

where the implied averages are over electron energy. The magnitude of $\beta$ depends on the details of the capture process.

When capture of an electron (or hole) leads to recombination with a hole (or electron), a recombination process has occurred. The lifetime of a free carrier, $\tau$, i.e., the average time the carrier is free before recombining, is given by

$$\tau = 1/\beta N \qquad (1.4)$$

Comparison with Eq. (1.2) shows that the rate of capture (in this case the rate of recombination) is equal to $n/\tau$. Since in steady state the rate of recombination must be equal to the rate of excitation $G$ (per

Figure 1.1. Major transitions and phenomena associated with photoelectronic effects in homogeneous semiconductors. (*a*) intrinsic absorption, (*b*) and (*c*) extrinsic absorption, (*d*) and (*e*) capture and recombination, (*f*) trapping and detrapping.

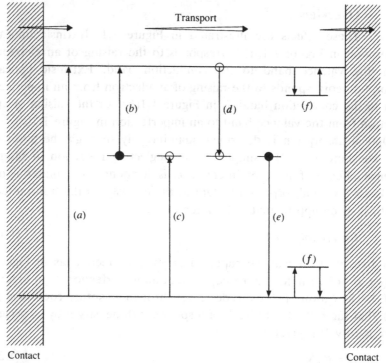

unit volume per second), we have the basic relationship that

$$n = G\tau \qquad (1.5)$$

This is a fundamental relationship in the discussion of all photo-electronic phenomena. It has the characteristic of a basic truism, applicable to many different types of situations, stating that the average density ($m^{-3}$) of a species present at a given time is given by the product of the rate at which the species is being generated ($m^{-3} s^{-1}$) multiplied by the average lifetime of a member of the species (s). If more than one type of recombination process is present, the individual recombination rates add.

A captured carrier at an imperfection may do one of two things: (i) recombine with a carrier of the opposite type, as just described, or (ii) be thermally reexcited to the nearest energy band before recombination occurs. In the latter case the imperfection is referred to as a trap, and the capture and release processes are called trapping and detrapping. Figure 1.1($f$) shows such a trapping and detrapping situation. If $R_c$ is the capture rate of free carriers with density $n$ by imperfections with density $N$, such that $R_c = \beta n N$, and if $R_d$ is the thermal detrapping rate given by $n_t v \exp(-\Delta E/kT)$, where $n_t$ is the density of trapped carriers, $v$ is a characteristic 'attempt to escape frequency,' and $\Delta E$ is the activation energy for detrapping, then the imperfection acts like a recombination center if $R_c > R_d$ but like a trap if $R_c < R_d$. We develop these concepts more completely later.

The physical situation of electrical current flow under illumination pictured in Figure 1.1 involves two other considerations: the effect of the contacts, and the nature of the transport of free carriers. Both of these must be considered in further detail.

The ability of the contacts to replenish carriers to maintain charge neutrality in the material (properties of an ohmic contact) if carriers are drawn out of the opposite contact by an electric field plays a key role in determining the kind of phenomena observed. The general effect of electrical contacts between a metal and a semiconductor under illumination can be described in terms of three classes of behavior: ($a$) blocking contacts, ($b$) ohmic contacts, and ($c$) injecting contacts. The characteristic dependence of current on applied voltage for these three types of contacts is shown in Figure 1.2. A qualitative description of the phenomena given below is followed by a more quantitative discussion in Section 1.4.

A blocking contact as in Figure 1.2($a$) is unable to replenish carriers created by photoexcitation when they are drawn out of the material by

Figure 1.2. Major variations of current with applied voltage for (*a*) a blocking contact, (*b*) an ohmic contact, and (*c*) an injecting contact.

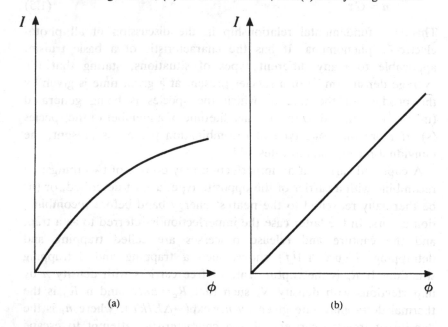

(a)                           (b)

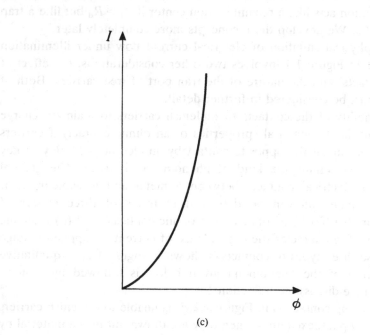

(c)

an applied electric field. The current varies less than linearly with applied voltage, and may become saturated.

An ohmic contact as in Figure 1.2(*b*) is able to replenish carriers created by photoexcitation when they are drawn out of the material by an applied electric field. The contacts contribute negligible electrical resistance, and current varies linearly with applied voltage with a slope equal to the reciprocal of the resistance of the bulk semiconductor.

An injecting contact as in Figure 1.2(*c*) is an ohmic contact at low applied electric fields that is being operated under sufficiently high applied electric field that carriers drawn in from the contact have time to cross the material in the electric field before the material is able to achieve charge neutrality by dielectric relaxation. The result is a regime in which the material is not electrically neutral, and the current flow is called 'space-charge-limited.'

The actual transport process, as usually described in terms of scattering effects or carrier mobility, determines how the change in free carrier density due to illumination affects the actual electrical conductivity. The total current density $\mathbf{J}_{tot}$ can be expressed as the sum of a drift current $\mathbf{J}_{dr}$ and a diffusion current $\mathbf{J}_{df}$:

$$\mathbf{J}_{tot} = \mathbf{J}_{dr} + \mathbf{J}_{df} \tag{1.6}$$

The drift current due to free electrons, for example, is given by

$$\mathbf{J}_{drn} = \sigma_n \mathscr{E} \tag{1.7}$$

where $\sigma_n$ is the electrical conductivity, $\sigma = nq\mu_n$ for conductivity due to electrons, with $n$ the density of free electrons, $q$ the charge per electron, and $\mu_n$ the electron mobility, and $\mathscr{E}$ is the electric field. The mobility can be written as

$$\mu_n = (q/m_n^*)\tau_{sc} \tag{1.8}$$

where $m_n^*$ is the effective mass of the electron and $\tau_{sc}$ is the scattering relaxation time for electrons, the average time between electron scattering events.

The electron drift current can also be written as

$$\mathbf{J}_{drn} = nq\mathbf{v}_{dn} \tag{1.9}$$

where $\mathbf{v}_{dn}$ is the electron drift velocity, such that

$$\mathbf{v}_{dn} = \mu_n \mathscr{E} \tag{1.10}$$

The electron diffusion current can be expressed as

$$\mathbf{J}_{dfn} = qD_n \nabla n \tag{1.11}$$

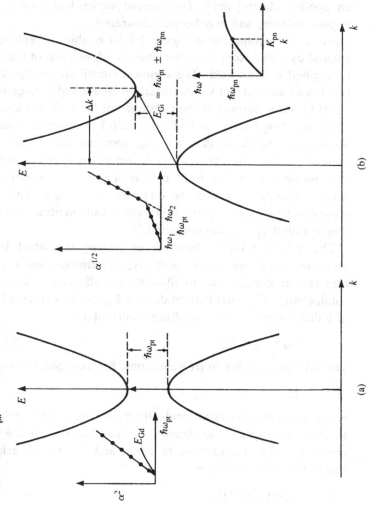

Figure 1.3. (a) Intrinsic direct absorption, and (b) intrinsic indirect absorption. Inserts show the variation of absorption constant $\alpha$ with photon energy $\hbar\omega_{\mathrm{pt}}$ expected for each type of absorption, and a typical phonon dispersion curve for the indirect material involving a phonon in the transition with energy $\hbar\omega_{\mathrm{pn}}$ and wavevector $K_{\mathrm{pn}}$.

where the electron diffusion coefficient $D_n$ is given by

$$D_n = \mu_n kT/q \tag{1.12}$$

## 1.2    Intrinsic optical absorption

Intrinsic optical absorption corresponds to photoexcitation of an electron from the valence band to the conduction band. If the minimum of the conduction band is at the same point in k-space as the maximum of the valence band, as shown in Figure 1.3(*a*), a vertical transition occurs involving only the absorption of a photon. Such a transition is called a direct optical transition. The minimum photon energy for absorption $\hbar\omega_{min} = E_{Gd}$, where $E_{Gd}$ is the direct bandgap of the material, and the change in $k$ upon making a transition $\Delta k = K_{photon} \approx 0$. This is because the momentum associated with the photon is very small (of the order of 1/500) compared to the width of the Brillouin zone and can be effectively neglected.

If the minimum of the conduction band is at a different point in k-space from the maximum of the valence band, as shown in Figure 1.3(*b*), an optical transition from the top of the valence band to the bottom of the conduction band must involve the absorption of a photon and a simultaneous absorption or emission of a phonon. In this case $E_{Gi} = \hbar\omega_{min} \pm E_{phonon}$, where $E_{Gi}$ is the indirect bandgap of the material and $E_{phonon}$ is the energy of the phonon involved in the process; a positive sign corresponds to phonon absorption, and a negative sign to phonon emission. The change in $k$ upon an indirect optical absorption is given by $\Delta k = K_{phonon}$.

To calculate the magnitude of the absorption constant $\alpha$ associated with an intrinsic optical transition, a quantum mechanical perturbation calculation must be carried out in which the effect of the light is treated as a perturbation in the basic Schroedinger equation.[1] For a direct transition, the transition probability depends only on the square of the matrix element involving the interaction of the light with the electrons, and is hence a first-order process. The final result is a variation of $\alpha$ for photon energies $\hbar\omega \geq E_{Gd}$ corresponding to

$$\alpha^2 = (\hbar\omega - E_{Gd}) \tag{1.13}$$

so that a plot of $\alpha^2$ vs $\hbar\omega$ gives a straight line with intercept of $E_{Gd}$, as shown in Figure 1.3(*a*). Direct optical transitions are characterized by a rapid increase in $\alpha$ for $\hbar\omega > E_{Gd}$ to values of the order of $10^6$–$10^7 \text{ m}^{-1}$, so that only a thin film of the material is required to absorb most of the light.

Quantum mechanical calculation of $\alpha$ for indirect optical transitions gives a transition probability that is proportional to the product of the square of the matrix element for photon–electron interaction and the square of the matrix element for phonon–electron interaction. The actual form of $\alpha$ is given by

$$\alpha \propto \langle n \rangle (\hbar\omega + E_{pn} - E_{Gi})^2 + (1 + \langle n \rangle)(\hbar\omega - E_{pn} - E_{Gi})^2 \tag{1.14}$$

where

$$\langle n \rangle = [\exp(E_{pn}/kT) - 1]^{-1} \tag{1.15}$$

is the Bose–Einstein distribution. The first term in Eq. (1.14) corresponds to optical absorption with absorption of a phonon, and the second to optical absorption with emission of a phonon. The magnitude of the $\alpha$s obtained are smaller than for direct absorption since they correspond to a second-order process. The magnitudes of $\alpha$ for photon energies slightly larger then $E_{Gi}$ are of the order of $10^4 \, \text{m}^{-1}$. A plot of $\alpha^{1/2}$ vs $\hbar\omega$, as shown in Figure 1.3($b$), consists of two straight lines, one corresponding to absorption of a phonon, and the other to emission of a phonon. If these two straight lines intersect the $\hbar\omega$ axis at $\hbar\omega_1$ and $\hbar\omega_2$ with $\hbar\omega_1 < \hbar\omega_2$, $E_{Gi} = (\hbar\omega_1 + \hbar\omega_2)/2$ and $E_{pn} = (\hbar\omega_2 - \hbar\omega_1)/2$.

A specific example is given in Figure 1.4 for the case of Ge at 300 K that has both a direct bandgap at $\mathbf{k} = 0$ and an indirect bandgap with conduction band minimum at the (111) zone face. $E_{Gi} = 0.68 \, \text{eV}$ and for $E < 0.8 \, \text{eV}$, the value of $\alpha$ is less than or equal to $10^4 \, \text{m}^{-1}$; $E_{Gd} = 0.81 \, \text{eV}$ and for $E > 0.8 \, \text{eV}$, the value of $\alpha$ is about $10^6 \, \text{m}^{-1}$. If the penetration depth $d$ is defined as $1/\alpha$, effectively the distance over which the intensity is reduced by a factor of $e$, then $d$ is about 100 $\mu$m for the indirect absorption range, and 1 $\mu$m for the direct absorption range.

Several critical effects dependent on the type of intrinsic optical excitation may be summarized as follows.

(1)  For a direct bandgap material (such as GaAs or CdTe) optical absorption occurs near the surface of a material; for an indirect bandgap material (such as Si or GaP) light penetrates much deeper into the material.

(2)  For direct bandgap materials, the magnitude of the intrinsic photoconductivity (defined as the increase in conductivity caused by intrinsic absorption) depends critically on the surface lifetime; in indirect materials the surface lifetime is much less important.

(3) If a direct bandgap material is used in a p–n junction type of device where junction collection of photoexcited carriers is important, a shorter diffusion length $L$ $(L = (D\tau)^{1/2})$ is required for carrier collection by the junction; in an indirect material a larger $L$ is necessary if all of the photoexcited carriers are to be collected.

(4) Direct bandgap materials have a higher intrinsic luminescence efficiency associated with a recombination of electrons and holes with emission of photons because of a shorter value of radiative recombination lifetime; indirect materials have a lower intrinsic luminescence efficiency, since the longer lifetime for radiative recombination allows competing processes for non-radiative recombination to become important.

## 1.3 Extrinsic optical absorption

Absorption involving imperfections is called extrinsic optical absorption. The absorption constant is proportional to the density of absorbing centers $N$,

$$\alpha = S_0 N \tag{1.16}$$

Figure 1.4. Dependence of absorption constant on photon energy for Ge. (After W. C. Dash and R. Newman, *Phys. Rev.* **99**, 1151 (1955).)

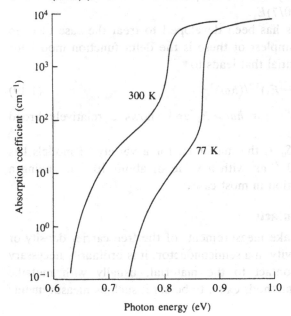

with a proportionality constant $S_0$ that is called the optical cross section for the absorption process. The magnitude and dependence on photon energy of $S_0$ have been calculated by a variety of quantum mechanical approximations.

Two extreme cases may be considered: (i) the case of an imperfection with a small ionization energy (a shallow imperfection) for which the uncertainty in position of a bound carrier $\Delta x$ is large, yielding a small uncertainty in $\Delta k$; and (ii) the other extreme of an imperfection with a large ionization energy (a deep imperfection) for which the wavefunction of the bound carrier is localized and the uncertainty $\Delta k$ is large.

A hydrogenic model can be applied to the case of a shallow imperfection in the electric dipole approximation, assuming a transition from a ground state like that of hydrogen to a final state consisting of a plane wave. The ionization energy of the imperfection $E_I$ can be expressed in terms of the ionization energy of the hydrogen atom $E_H$, the effective mass of the free carriers $m^*$, the free carrier mass $m_0$ and the dielectric constant of the material $\epsilon$, as

$$E_I = E_H[(m^*/m_0)/\epsilon^2] \tag{1.17}$$

and the dependence of $S_0$ on photon energy is given by

$$S_0(\hbar\omega) \propto (\hbar\omega - E_I)^{3/2}/(\hbar\omega)^5 \tag{1.18}$$

which increases rapidly for $\hbar\omega > E_I$ and shows a relatively sharp maximum for $\hbar\omega = (10/7)E_I$.

A variety of models has been developed to treat the case of deep imperfections.[2] The simplest of these is the delta function model for the imperfection potential that leads to

$$S_0(\hbar\omega) \propto (\hbar\omega - E_I)^{3/2}/(\hbar\omega)^3 \tag{1.19}$$

which increases rapidly for $\hbar\omega > E_I$ and shows a relatively broad maximum for $\hbar\omega = 2E_I$.

The magnitude of $S_0$ at the maximum for a variety of models lies between $10^{-19}$ and $10^{-21}\,m^2$ with a value of about $10^{-20}\,m^2$ being a reasonable approximation in most cases.

## 1.4    Electrical contacts

In order to make measurements of the free carrier density or the electrical conductivity in a semiconductor, it is ordinarily necessary to make electrical contact to the material, usually with metallic contacts. Alternative methods exist, to be sure, such as measurements of free carrier absorption or the time constants associated with the charging and discharging of thin films of the material. Both blocking and ohmic contacts can be used to obtain such information.

If the effects of surface states can be neglected, then the simplest criterion for the behavior of contacts is the relative magnitude of the metal workfunction $q\phi_M$ and the semiconductor workfunction $q\phi_s$ (the energy difference between the vacuum level and the Fermi level in the material). For an n-type semiconductor, a blocking contact is obtained if $q\phi_M > q\phi_s$ (electrons transferred from the semiconductor to the metal upon contact, leaving a depletion layer in the semiconductor), and an ohmic contact is obtained if $q\phi_M < q\phi_s$ (electrons transferred from the metal to the semiconductor upon contact, producing an accumulation layer in the semiconductor). The relationships are simply reversed for a p-type semiconductor.

### Blocking contacts

A typical blocking contact to an n-type semiconductor is shown in Figure 1.5. Such a contact is also called a rectifying contact, a non-ohmic contact, or a Schottky barrier. The quantity $\phi_D$ shown in the figure is called the diffusion potential and $q\phi_D$ is equal to the difference between the workfunctions of the metal and the semicon-

Figure 1.5. A Schottky barrier blocking contact between a metal and an n-type semiconductor, formed when the workfunction of the metal is larger than that of the semiconductor.

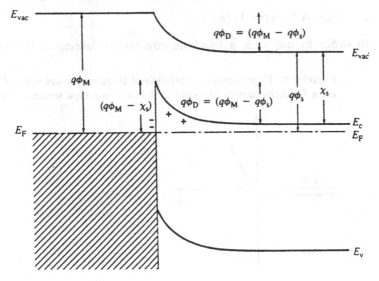

ductor. The barrier height $V_b$ faced by an electron moving from the metal to the semiconductor is larger than this by the amount $(E_c - E_F)$, where $E_c$ is the energy of the bottom of the conduction band in the semiconductor, and $E_F$ is the Fermi energy in the semiconductor. If the concept of the electron affinity of the semiconductor $\chi_s$ is introduced, then this barrier height can be expressed as $V_b = (q\phi_M - \chi_s)$.

The depletion layer in the n-type semiconductor with width $w_d$ is assumed to be devoid of electrons, containing only the charge of the ionized donors $N_{D^+}$. In a more general situation, the charge in $w_d$ is the net uncompensated positive charge $(N_{D^+} - N_{A^-})$, if $N_{A^-}$ ionized acceptors are also present, which can change under illumination.

If a voltage difference is applied across such a blocking contact, it is evident that the observed current will depend critically on the polarity of this applied voltage. Figure 1.6 shows the typical case for an n-type semiconductor. If the metal is positive, the energy levels in the semiconductor are raised, more electrons flow into the metal and the current increases exponentially with applied voltage $\phi_a$; this is called the forward bias direction. If the metal is negative, the barrier height $V_b$ is unaffected, and a small current independent of applied voltage is observed. The voltage dependence of the current can be expressed as

$$\mathbf{J} = \mathbf{J}_0[\exp(q\phi_a/kT) - 1] \tag{1.20}$$

Assuming that current flows across the metal–semiconductor interface by thermionic emission in both directions leads to a form for $J_0$ of

$$J_0 = AT^2 \exp(-V_b/kT) \tag{1.21}$$

In order to use such a blocking contact to determine the carrier

Figure 1.6. Representative variation of current with applied voltage for a Schottky barrier blocking contact on an n-type semiconductor.

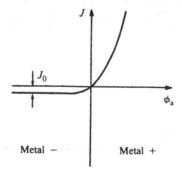

Metal −          Metal +

density in the semiconductor, something more than a simple measurement of current vs voltage is required. The fact that the metal/depletion layer/semiconductor combination corresponds to a capacitor with the depletion layer as the dielectric leads to the possibility of measuring the capacitance of this structure, and from the capacitance being able to deduce in a simple case both the diffusion potential of the junction and the value of $N_{D^+}$ which can be assumed to be equal to the free electron density $n$ in the bulk semiconductor.

Poisson's equation for the depletion region is

$$\partial^2\phi/\partial x^2 = -N_{D^+}q/\epsilon_r\epsilon_0 \tag{1.22}$$

if we assume that $N_{D^+}$ is the positive charge in the depletion region. If $x = 0$ at the junction interface and $x = w_d$ at the edge of the depletion layer, then the boundary conditions applicable to the solution of this equation are $\mathscr{E} = \partial\phi/\partial x = 0$ at $x = w_d$, and $\phi = \phi_D$ at $x = w_d$. Integration of Eq. (1.22) yields

$$w_d = (2\epsilon_r\epsilon_0\phi_D/qN_{D^+})^{1/2} \tag{1.23}$$

If an applied voltage $\phi_a$ is also present,

$$w_d = [2\epsilon_r\epsilon_0(\phi_D - \phi_a)/qN_{D^+}]^{1/2} \tag{1.24}$$

Since the capacitance of the junction is given by

$$C = \epsilon_r\epsilon_0 A/w_d \tag{1.25}$$

it follows that

$$(C/A)^{-2} = 2(\phi_D - \phi_a)/(\epsilon_r\epsilon_0 qN_{D^+}) \tag{1.26}$$

A plot of $(C/A)^{-2}$ vs $\phi_a$ then yields a straight line with intercept of $\phi_D$ and slope proportional to $1/N_{D^+}$.

The direct correlation between the measured capacitance and the uncompensated charge in the depletion region provides a useful approach for several different kinds of photoelectronic analysis of imperfections in semiconductors.

Low-resistance quasi-ohmic contacts to an n-type material can be produced even in the case of metals and semiconductors where $q\phi_M > q\phi_s$ by heavily n-type doping the surface of the material to which contact is being made (e.g., either by diffusion of donors or ion implantation of donors) so that it becomes possible to tunnel through the barrier rather than being thermally excited over it. Doping the semiconductor near the surface reduces the width of the depletion layer and hence the distance over which tunneling must occur.

### Ohmic contacts

A typical ohmic contact to an n-type semiconductor is shown in Figure 1.7. For low applied voltages, such that the current flow is controlled by the resistance of the bulk semiconductor, a linear current vs voltage curve is obtained. For higher applied voltages it becomes possible for carriers injected from the contact to dominate the current flow, as discussed in the following section.

The variation of potential energy with distance in the contact/accumulation layer region can be expressed by equating drift and diffusion currents:

$$n(x)q\mu_n\mathscr{E} = q(\mu_n kT/q)\,dn(x)/dx \tag{1.27}$$

Integration of this equation gives

$$n(x) = n_0 \exp[-q\phi(x)/kT] \tag{1.28}$$

since $\mathscr{E} = -\nabla\phi(x)$, where $n_0$ is the electron density at the junction interface ($x = 0$). Now Poisson's equation is

$$\partial^2\phi/\partial x^2 = -n_0 q\,\exp(-q\phi(x)/kT)/\epsilon_r\epsilon_0 \tag{1.29}$$

with solution

$$q\phi(x) = 2kT\,\ln[(x/x_0) + 1] \tag{1.30}$$

Figure 1.7. An ohmic contact between a metal and an n-type semiconductor, formed when the workfunction of the metal is smaller than that of the semiconductor.

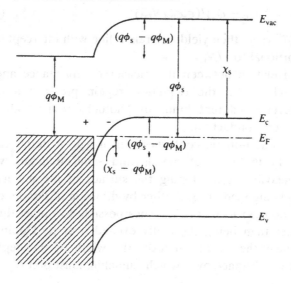

with

$$x_0 = (2\epsilon_r\epsilon_0 kT/n_0 q^2)^{1/2} \tag{1.31}$$

and

$$n(x) = n_0[x_0/(x + x_0)]^2 \tag{1.32}$$

Eq. (1.32) describes the spatial dependence of the available reservoir of excess electrons present in the semiconductor.

### *Single injection*

When the voltage applied to ohmic contacts becomes sufficiently large, the carriers injected from the contact may dominate the measured current flow. A representative energy-level diagram is shown in Figure 1.8. A simple conceptual model can be stated as follows. If the spacing between electrodes is $d$ and this is treated as the dielectric between two capacitor plates with capacitance $C$, then the

Figure 1.8. Energy-level diagrams for a material with ohmic contacts for electrons; ($a$) in the absence of an applied field; and ($b$) in the presence of an applied field, showing how the accumulation layer behaves like a kind of virtual cathode.

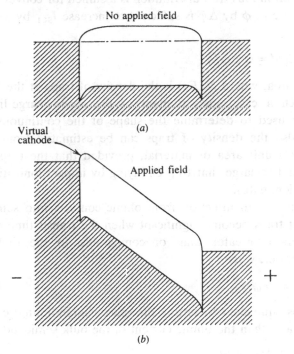

charge $Q$ per unit area induced by a voltage $\phi$ is

$$Q = C\phi = \epsilon_r\epsilon_0\phi/d \tag{1.33}$$

The transit time $t_{tr}$ for charge to pass through the material is

$$t_{tr} = d/\mu\mathscr{E} = d/(\mu\phi/d) = d^2/\mu\phi \tag{1.34}$$

An approximation to the space-charge-limited current density $J_{SCL}$ is given by

$$J_{SCL} = Q/t_{tr} = \epsilon_r\epsilon_0\mu\phi^2/d^3 \tag{1.35}$$

Therefore, for a semiconductor without traps in which all of the injected charge remains free, the space-charge-limited current density varies as $\phi^2/d^3$. These currents can be large; if 10 V is applied across $10^{-5}$ m of a trap-free material with $\mu$ of $10^{-2}$ m$^2$ V$^{-1}$ s$^{-1}$ and $\epsilon_r = 10$, $J_{SCL}$ is about $10^5$ A m$^{-2}$.

If the semiconductor contains traps, then the measured value of $J_{SCL}$ is much smaller for lower voltages, as the injected charge fills traps. When the applied voltage is sufficiently large that all of the traps are filled, $\phi = \phi_{TF}$, the current rises rapidly to the trap-free limit described by Eq. (1.35). Over none of the range is a simple variation as $\phi^2$ expected. If a continuous trap distribution is assumed for convenience, then if an increase in $\phi$ by $\Delta\phi$ is sufficient to increase $J_{SCL}$ by a factor of $e$,

$$kTN_t(E_F)d = C\,\Delta\phi/q \tag{1.36}$$

for $C$ per unit area, where $N_t(E_F)$ is the density of traps at the Fermi energy. In such a case, measurement of the space-charge-limited current can be used to determine the shape of the continuous trap distribution. Also the density of traps can be estimated from $N_t = \epsilon_r\epsilon_0\phi_{TF}/qd$ for a unit area of material, provided that the trap-filled field $\phi_{TF}/d$ is not so large that trap emptying by impact ionization or field emission dominates.

Such currents due to injection from ohmic contacts into semiconductors without traps become significant when the transit time of the injected carriers is greater than or comparable to the dielectric relaxation time, i.e., when

$$d^2/\mu\phi \gtrsim \epsilon_r\epsilon_0/\sigma \tag{1.37}$$

This criterion is equivalent to setting the space-charge-limited current equal to or greater than the ohmic current of the bulk semiconductor,

$$\epsilon_r\epsilon_0\mu\phi^2/d^3 \geq \sigma\phi/d \tag{1.38}$$

This gives a value for the critical voltage onset of space-charge-limited currents $\phi_{crit}$ as

$$\phi_{crit} = \sigma d^2 / \epsilon_r \epsilon_0 \mu \tag{1.39}$$

For typical values of $\sigma = 10^{-4} \, \Omega^{-1} \, m^{-1}$, $d = 10^{-3} \, m$, $\epsilon_r = 10$, and $\mu = 10^{-1} \, m^2 \, V^{-1} \, s^{-1}$, a value of $\phi_{crit} = 11 \, V$ is obtained for this trap-free example.

If the calculation resulting in Eq. (1.35) is considered too intuitive, more accurate calculations of the problem can be carried out with substantially the same result.[3] For the case of an insulator, for example, solution for $n$ from $\mathbf{J} = nq\mu\mathscr{E}$ and substitution into $d\mathscr{E}/dx = nq/\epsilon_r\epsilon_0$, yields with boundary condition that $\mathscr{E}(0) = 0$, $\mathscr{E} = -d\phi/dx$, and $\phi = \phi(d)$ at $x = d$ that $J_{SCL} = \frac{9}{8}(\epsilon_r\epsilon_0\mu\phi^2/d^3)$, identical to Eq. (1.35) except for the factor of $\frac{9}{8}$.

### Double injection

Under appropriate conditions it may develop that electrons are injected from one contact while holes are injected from the other. In the material the space charge will be partially cancelled by the presence of carriers of both sign. It is expected, therefore, that overall more current will flow under double injection than single injection conditions. A simple model that produces the correct functional relationships can be developed. Let $Q^*$ be the total uncompensated injected charge, determined by how far the injected carriers go before recombination. If $n = p$,

$$Q^* = C\phi = Q_n^* + Q_p^* \tag{1.40}$$

which will be given approximately by

$$Q^* \approx nq[t_n/\tau_n + t_p/\tau_p] \tag{1.41}$$

where $t_i$ are the transit times, and $\tau_i$ are the lifetimes. From Eq. (1.41) with $\tau_n = \tau_p$, we obtain per unit area,

$$t_n + t_p = \epsilon_r\epsilon_0\tau\phi/qnd \tag{1.42}$$

Now the space-charge-limited current can be expressed as

$$J_{SCL} = Q_n/t_n + Q_p/t_p = nq(1/t_n + 1/t_p) = nq(t_n + t_p)/t_n t_p \tag{1.43}$$

Using $(t_n + t_p)$ from Eq. (1.42), $t_n = d^2/\mu_n\phi$, and $t_p = d^2/\mu_p\phi$, gives

$$J_{SCL} = \mu_n\mu_p\tau\epsilon_r\epsilon_0\phi^3/d^5 \tag{1.44}$$

In this trap-free case, therefore, $J$ is expected to vary as $\phi^3/d^5$ for double injection conditions.

# 2

# Photoconductivity parameters

The description of photoconductivity phenomena focuses on the definitions of three basic quantities: the photosensitivity, the spectral response, and the speed of response. In this chapter we consider these quantities, take a look at the simple situation of photoconductivity kinetics in an intrinsic material, and describe differences in several different types of basic photoconducting systems.

## 2.1    Overview

In a semiconductor with free electrons and/or free holes, the electrical conductivity in the dark is given by

$$\sigma_0 = q(n_0\mu_{n0} + p_0\mu_{p0}) \tag{2.1}$$

In the light, this conductivity is increased by the photoconductivity $\Delta\sigma$,

$$\sigma_L = \sigma_0 + \Delta\sigma \tag{2.2}$$

Consider the case of one-carrier transport for simplicity. Then in the dark $\sigma_0 = n_0 q \mu_0$ and in the light

$$\sigma_0 + \Delta\sigma = (n_0 + \Delta n)q(\mu_0 + \Delta\mu) \tag{2.3}$$

allowing for the possibility that photoexcitation may change both the carrier density $\Delta n$ and the carrier mobility $\Delta\mu$. Then

$$\Delta\sigma = q\mu_0 \Delta n + (n_0 + \Delta n)q \Delta\mu \tag{2.4}$$

Now it is generally true that

$$\Delta n = G\tau_n \tag{2.5}$$

where $G$ is the photoexcitation rate $(m^{-3} s^{-1})$ and $\tau_n$ is the electron lifetime, so that

$$\Delta\sigma = q\mu_0 G\tau_n + nq\,\Delta\mu \qquad (2.6)$$

One additional complexity arises from the fact that the lifetime $\tau_n$ may itself be a function of excitation rate $\tau_n(G)$. There are therefore three types of effects that may manifest themselves in Eq. (2.6).

(1) Increase in carrier density with constant lifetime $\tau_n$, so that

$$\Delta\sigma = q\mu_0\tau_n G \qquad (2.7)$$

The photoconductivity is proportional to $G$ so that a log–log plot of $\Delta\sigma$ vs $G$ shows a slope of 1.

(2) Increase in carrier density with lifetime $\tau_n$ a function of photoexcitation intensity, so that

$$\Delta\sigma = q\mu_0\tau_n(G)G \qquad (2.8)$$

If $\tau_n$ varies as $G^{(\gamma-1)}$, the $\Delta\sigma$ varies as $G^\gamma$. If $\gamma < 1$, the lifetime decreases with increasing excitation rate; the behavior is said to be 'sublinear.' If $\gamma > 1$, the lifetime increases with increasing excitation rate; the behavior is said to be 'supralinear.'

(3) Increase in carrier mobility, so that

$$\Delta\sigma = nq\,\Delta\mu \qquad (2.9)$$

A number of possible mechanisms may give rise to this kind of behavior: (*a*) scattering by charged impurities may change under photoexcitation either through a change in the density of such charged impurities or through a change in the scattering cross section of such impurities; (*b*) if the material is polycrystalline and contains intergrain potential barriers, photo-excitation can reduce the height of these barriers as well as the depletion width in the grains adjacent, giving rise to an increase in carrier mobility; (*c*) photoexcitation may excite a carrier from a band characterized by one mobility to a band characterized by a different mobility.

In the case of insulators at reasonably high photoexcitation rates, $\Delta n \gg n_0$, $\Delta p \gg p_0$, and

$$\Delta\sigma \approx Gq(\tau_n\mu_n + \tau_p\mu_p) \qquad (2.10)$$

We may define a 'figure of merit' for a single-carrier photoconductor as

$$\Delta\sigma/Gq = \tau\mu \qquad (2.11)$$

Thus the 'lifetime–mobility' product is a measure of the photoconductor's sensitivity to photoexcitation. There are other ways in which such photosensitivity is described. We compare these next.

## 2.2  Measures of photosensitivity

Two major definitions for 'photosensitivity' are found in the literature that must be carefully distinguished. One of these simply describes photosensitivity by the magnitude of $\Delta\sigma$, whereas the other, more device-oriented, definition is in terms of the ratio $\Delta\sigma/\sigma_0$. These two definitions may often be contradictory, as, for example, in the case of a semiconductor in which both $\Delta\sigma$ and $\sigma_0$ increase with a particular processing, but $\sigma_0$ increases more rapidly than $\Delta\sigma$. Then after processing, the 'photosensitivity' defined as $\Delta\sigma$ is increased, but the 'photosensitivity' defined as $\Delta\sigma/\sigma_0$ is decreased. Following are several ways in which photosensitivity may be defined.

*Specific sensitivity*

The quantity called 'specific sensitivity' is a measure of the material's actual sensitivity in terms of a $\tau\mu$ product. It is usually cited for large-signal photoconductivity ($\Delta\sigma \gg \sigma_0$) in insulators.

$$S^* = \Delta I\, d^2/\phi P \text{ m}^2\,\Omega^{-1}\,\text{W}^{-1} \tag{2.12}$$

where $P$ is the input light power. Setting $\Delta I/\phi = \Delta\sigma A/d$, and $P = Gh\nu dA$, an analysis of $S^*$ shows that

$$S^* = (q/h\nu)\tau\mu \tag{2.13}$$

For $h\nu = 2\,\text{eV}$, $\tau = 10^{-3}\,\text{s}$, and $\mu = 10^{-2}\,\text{m}^2\,\text{V}^{-1}\,\text{s}^{-1}$, $S^* = 5 \times 10^{-6}\,\text{m}^2\,\Omega^{-1}\,\text{W}^{-1}$.

*Detectivity*

The measure of photosensitivity known as 'detectivity' is usually applied to the small-signal ac-$\Delta\sigma$ situation where $\Delta\sigma \leq \sigma_0$, most commonly in applications like that of an infrared detector, where it is desired to detect small photoexcited signals in the presence of background noise.

$$D^* = (A\,\Delta\nu)^{1/2}NEP \tag{2.14}$$

where $A$ is the area of the detector, $\Delta\nu$ is the bandwidth used for measurement of noise, and $NEP$ represents the radiation power needed to produce a signal equal to the noise. Typical values of $D^*$ lie in the range between $10^9$ and $10^{11}$.

*Photoconductivity gain*

A third measure of photosensitivity is the combination of specific sensitivity and device parameters known as photoconductivity gain. Specifically the gain of a photodetector can be defined as the number of charges collected in the external circuit for each photon absorbed. If $F$ is the total number of photons absorbed, so that the photoexcitation rate $G = F/Ad$, then the gain $G^*$ can be expressed:

$$G^* = (\Delta I/q)/F = \tau\mu\phi/d^2 \tag{2.15}$$

by setting $\Delta I = \Delta\sigma A\phi/d$, $\Delta\sigma = \Delta nq\mu$, and $\Delta n = G\tau$.

Alternatively we can view this result as being equivalent to saying that the gain of a photoconductor is equal to the ratio of the carrier lifetime to the carrier transit time, i.e., $G^* = \tau/(d^2/\mu\phi)$. Physically we may view the process as follows. Photoexcitation creates free electrons and holes. If the lifetime of a carrier is greater than its transit time, it will make several effective transits through the material between the contacts, provided that the contacts are ohmic and are able to replenish carriers drawn off at the opposite contact.

If both electrons and holes contribute to the photoconductivity, then

$$G^* = (\tau_n\mu_n + \tau_p\mu_p)\phi/d^2 = (h\nu/q)(\phi/d^2)S^* \tag{2.16}$$

If $h\nu = 2\,\text{eV}$, $\tau = 10^{-3}\,\text{s}$, $\mu = 10^{-2}\,\text{m}^2\,\text{V}^{-1}\,\text{s}^{-1}$, $\phi = 10^2\,\text{V}$, and $d = 10^{-3}\,\text{m}$, the gain $G^* = 10^3$.

## 2.3 Spectral response of photoconductivity

In order for photoconductivity to be produced, light must be absorbed in the process of creating free carriers by either intrinsic or extrinsic optical absorption. There is, therefore, a close correlation between the optical absorption spectrum $\alpha$ vs $\hbar\omega$ and the photoconductivity spectral response $\Delta\sigma$ vs $\hbar\omega$.

Typical spectral response of photoconductivity curves are given in Figure 2.1($a$); the general shape of these curves is pictured in Figure 2.1($b$). In the high absorption region I, the photoconductivity is controlled by the surface lifetime. In the intermediate range of region II, there is still strong absorption and the photoconductivity is controlled by the bulk lifetime, with a maximum occurring when the absorption constant is approximately equal to the reciprocal of the sample thickness. In the low absorption region III, the photoconductivity is also controlled by the bulk lifetime, but decreases with increasing wavelength as the absorption decreases.

A common geometry for spectral response measurements is for the

Figure 2.1. (*a*) Photoconductivity spectral response curves for a photosensitive crystal of CdS at 90 and 300 K.[1] (*b*) Illustrative comparison of the shape of the photoconductivity spectral response and the absorption constant of the material, defining three characteristic regions. The absorption edge for CdS at 300 K is about 5150 Å.

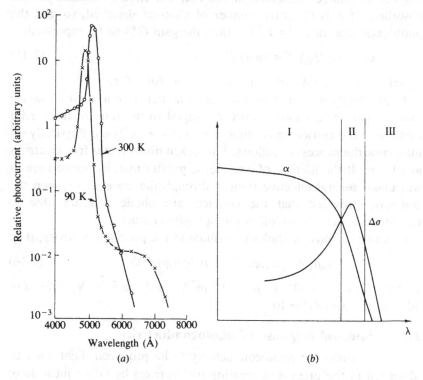

(*a*)                                                    (*b*)

Figure 2.2. Geometry for the calculation of the photoconductivity spectral response when the excitation is incident normal to the electric field direction.

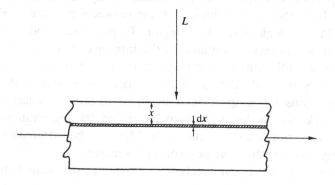

light to be incident normal to the applied electric field, as indicated in Figure 2.2. In this case the photoconductivity spectral response can be fairly accurately calculated from a simple model.[2] At a distance $x$ from the surface, the rate of free carrier generation can be expressed as

$$\mathrm{d}n(x)/\mathrm{d}t = A \exp(-\alpha x) - \mathrm{d}I/\mathrm{d}x - n/\tau \tag{2.17}$$

It is assumed in this calculation that the photoexcited carrier density is sufficiently larger than the dark carrier density that for all practical purposes $\Delta n = n$. The first term on the right of Eq. (2.17) represents the rate of optical generation of free carriers, the second represents the diffusion of free carriers in the $x$-direction with $I = n \,\mathrm{d}x/\mathrm{d}t = -D \,\mathrm{d}n/\mathrm{d}x$ with $D = \mu kT/q$, and the third represents recombination of free carriers with a bulk lifetime that is independent of photoexcitation rate. If $L$ (m$^{-2}$) is the total light absorbed,

$$L = \int_0^\infty A \exp(-\alpha x) \,\mathrm{d}x = A/\alpha \tag{2.18}$$

so that $A = \alpha L$. Then in steady state with $\mathrm{d}n(x)/\mathrm{d}t = 0$,

$$\mathrm{d}^2 n(x)/\mathrm{d}x^2 = n(x)/D\tau - (\alpha L/D) \exp(-\alpha x) \tag{2.19}$$

The general solution of this equation is

$$n(x) = C_1 \exp(x/L_D) + C_2 \exp(-x/L_D) + [L\alpha\tau/(1 - \alpha^2 D\tau)] \exp(-\alpha x) \tag{2.20}$$

where $L_D$ is the carrier diffusion length, $L_D = (D\tau)^{1/2}$.

The applicable boundary conditions may be expressed:

$$x = 0 \qquad I = -n(0)s = -D(\mathrm{d}n/\mathrm{d}x)|_{x=0} \tag{2.21}$$

$$x = d \qquad I = n(d)s = -D(\mathrm{d}n/\mathrm{d}x)|_{x=d} \tag{2.22}$$

where $s$ is the surface recombination velocity for the top and bottom surfaces. The surface recombination rate is proportional to the carrier density at the surface, and $s$ (m s$^{-1}$) is the proportionality constant. The complete solution finally involves calculating $n = \int_0^d n(x) \,\mathrm{d}x$.

The result can be expressed as follows:

$$n/L\tau = \{[1 - \exp(-Z)]/[1 + R \coth(W/2)]\}$$
$$\times \{1 + RW[W \coth(W/2) - Z \coth(Z/2)]/(W^2 - Z^2)\} \tag{2.23}$$

in terms of three dimensionless parameters: a thickness parameter $W$, given by

$$W = d/L_D \tag{2.24a}$$

a recombination parameter $R$, given by

$$R = s(\tau/D)^{1/2} \tag{2.24b}$$

and an absorption parameter $Z$, given by

$$Z = \alpha d \tag{2.24c}$$

Limiting cases of Eq. (2.23) are easy to define and recognize. If $\alpha \to 0$, i.e., if there is no absorption, then $Z \to 0$ and $n \to 0$. If, on the other hand, $\alpha \to \infty$, all of the photoexcitation occurs at the surface, $Z \to \infty$, and $n \to$ constant independent of wavelength. If $W \to 0$, $d \ll L_D$, the surface dominates, and there is no maximum in $n$. If $R \to 0$, $s \ll (D/\tau)^{1/2}$, the bulk dominates, and there is no maximum in $n$. It is only in the case that $s \gg (D/\tau)^{1/2}$ that a maximum occurs when there is a transfer of dominance of recombination by the surface at shorter wavelengths to dominance of recombination by the bulk at longer wavelengths corresponding to $\tau_{surf} < \tau_{bulk}$.

Many problems in photoelectronic effects involve a calculation of the photoexcitation rate as in Eq. (2.18). There are two extreme cases: (a) when $\alpha \ll 1/d$, where $d$ is the thickness of the material, the photoexcitation rate is approximately constant throughout the material, or (b) when $\alpha \gg 1/d$, the photoexcitation is concentrated in a thickness of the material equal to the penetration depth $d^* = 1/\alpha$. Let us express the average photoexcitation rate over a thickness of material $w$ by

$$G_{avg} = (1/w) \int_0^w \alpha L \exp(-\alpha x)\, dx = (L/w)[1 - \exp(-\alpha w)] \tag{2.25}$$

For small values of $\alpha \ll 1/d$, where $d$ is the thickness of the material, the effective $w = d$, and $G_{avg} = (L/d)(\alpha d) = \alpha L$. For the intermediate case where $\alpha = 1/d$, $G_{avg} = (L/d)(1 - 1/e) = 0.63\,\alpha L$. For large values of $\alpha \gg 1/d$, the effective $w = d^*$, $G_{avg} = L/d^* = \alpha L$. So for the two extreme cases, one can consider $G_{avg} = \alpha L$ for both, provided that one recognizes that in the case of small $\alpha$, one is talking about the whole material thickness, whereas in the case of large $\alpha$, one is taking only about a thickness of the material equal to the penetration depth.

The above calculation was for the case in which the photoexcitation direction was normal to that of the applied electric field. A model for the related case where the photoexcitation direction is parallel to the applied field has also been developed,[3] which considers only drift currents and neglects diffusion effects. This starts once again with the generation relationship of $dn(x)/dt = \alpha L \exp(-\alpha x)$, but now must take account of the actual polarity of the applied voltage, since if the illuminated face is positive, photoexcited electrons will be drawn to it while photoexcited holes will be drawn to the bottom contact; if the illuminated face is negative, the opposite behavior of the photoexcited carriers occurs.

For the illuminated face positive, for example, there is a contribution due to electron current,

$$dI_n = (qw_n/d)[1 - \exp(-x/w_n)](dn/dt)dx \qquad (2.26)$$

and one due to hole current,

$$dI_p = (qw_p/d)\{1 - \exp[-(d-x)/w_p]\}(dp/dt)\,dx \qquad (2.27)$$

where $x$ is measured from the illuminated contact from 0 to $d$, and where $w_n = \mu_n \mathscr{E} \tau_n$ and $w_p = \mu_p \mathscr{E} \tau_p$ are sometimes called the electron and hole schubweg, respectively, and $d$ is the distance between contacts. The total measured current is given by $J_+ = \int_0^d (dI_n + dI_p)$. Similar expressions hold for $J_-$, the measured current when the illuminated contact is negative.

The results can be expressed in terms of three non-dimensional quantities: $b = \mu_n \tau_n / \mu_p \tau_p$, $\beta = d^2/\mu_n \tau_n \phi$, and $K = \alpha d$.

$$
\begin{aligned}
J_+ = qL(&(b+1)[1 - \exp(-K)]/\beta b \\
&+ K\{\exp[-(K+\beta)] - 1\}/[\beta(K+\beta)] \\
&+ K[\exp(-K) - \exp(-\beta b)]/[\beta b(K - \beta b)])
\end{aligned}
\qquad (2.28)
$$

$$
\begin{aligned}
J_- = qL(&(b+1)[1 - \exp(-K)]/\beta b \\
&+ K\{\exp[-(K+\beta b)] - 1\}/[\beta b(K + \beta b)] \\
&+ K[\exp(-K) - \exp(-\beta)]/[\beta(K - \beta)])
\end{aligned}
\qquad (2.29)
$$

These results can also produce a maximum in the spectral response; such a maximum is expected, for example, in $J_+$ if $b \gg 1$. Experimental results for photoconductivity spectral responses measured with this geometry are given in Figure 2.3. The description given by Eqs. (2.28) and (2.29) does not completely describe the observed results, e.g., the magnitude of the ratio $J_{-max}/J_{+max}$ is larger than predicted, and the decrease in $J_-$ for shorter wavelengths is much larger than predicted. It is probable that the incorporation of diffusion of minority carriers to the surface with recombination there with majority carriers being collected at the surface, is needed to accurately treat the high-absorption conditions.

## 2.4    Speed of response: effect of traps

The 'speed of response' is inversely proportional to the time constant associated with the increase of photoconductivity to its steady state value after turning on the photoexcitation (the rise time), and the time constant associated with the decrease of photoconductivity to its dark value after turning off the photoexcitation (the decay time). In the absence of traps in the material the free carrier lifetime determines the value of these time constants. The simplest rate equation (assuming $\Delta n = n$) is

$$dn/dt = G - n/\tau \tag{2.30}$$

with steady state result $n = G\tau$. The rise curve is described by integrating Eq. (2.30) to obtain

$$n(t) = G\tau[1 - \exp(-t/\tau)] \tag{2.31}$$

Figure 2.3. Spectral response of photocurrent density for a crystal of HgI$_2$ with transverse geometry in which light and electric field are parallel.[4] $J_-$ is the photocurrent density when the illuminated surface is negative, and $J_+$ is the photocurrent density when the illuminated surface is positive.

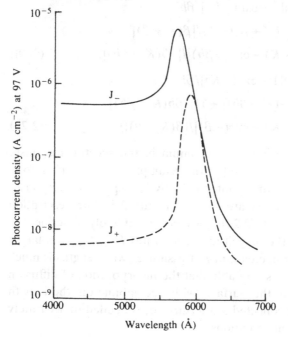

and the decay curve is obtained by integrating Eq. (2.30) with $G = 0$,

$$n(t) = G\tau \exp(-t/\tau) \tag{2.32}$$

In a later section we consider in some detail the use of photoconductivity transients in photoelectronic analysis. Here we wish to establish three major effects of traps in a semiconductor: (*a*) a decrease in the speed of response, (*b*) a decrease in the drift mobility, and (*c*) a decrease in the photosensitivity.

### Decrease in speed of response

In the presence of traps, additional time-dependent processes are involved with trap filling during the rise and trap emptying during the decay. Therefore the measured response time $\tau_0$, which we may take to be the time required for the photoconductivity to decay to $1/e$ of $G\tau$, can never be smaller than the lifetime $\tau$, but can be much larger. The ratio of $\tau_0/\tau$ is usually a strong function of the excitation rate $G$, since if $G$ is large enough the density of free carriers can be made much larger than the density of trapped carriers and $\tau_0 \to \tau$. This statement can be given quantitative form by writing

$$\tau_0 = \tau(1 + n_t/n) \tag{2.33}$$

where $n_t$ is defined to be the density of traps that must empty in order for the steady state Fermi energy to drop by $kT$ (the requirement that the photoconductivity decrease by $1/e$). If $n \gg n_t$, as is the case for low trap densities and/or high photoexcitation rates, $\tau_0 = \tau$, whereas if $n \ll n_t$, as is the case for high trap densities and/or low photoexcitation rates, $\tau_0 = \tau n_t/n$.

### Decrease in drift mobility

The drift mobility is determined by the time it takes carriers injected at one point in the sample to travel to a second point distant from it by $d$ under an electric field $\mathscr{E}$: $\mu_d = d/\mathscr{E}t$. If there are no traps, the injected carriers are free and the drift mobility is the same as the conductivity mobility $\mu$. If traps are present, however, we may write $\mu_d = \gamma\mu$, where $\gamma$ is the average time spent by carriers in the conduction band between traps, divided by the average time required to escape from a trap. Therefore

$$\gamma = (1/vS_tN_t)/\{1/N_cvS_t \exp[-(E_c - E_t)/kT]\}$$

$$= (N_c/N_t) \exp[-(E_c - E_t)/kT] \tag{2.34}$$

where $S_t$ is the trap capture cross section, $N_t$ is the trap density, $N_c$ is the effective density of states in the band, and $(E_c - E_t)$ is the trap depth.

An alternative perspective is to recognize that the product of all of the injected carriers multiplied by the drift mobility is equal to the product of the free carriers multiplied by the conductivity mobility:

$$(n + n_t)\mu_d = n\mu \qquad (2.35)$$

where $n$ is the density of injected free carriers and $n_t$ is the density of injected trapped carriers. Eq. (2.35) is equivalent to Eq. (2.34) when $n \ll n_t$, since $n = N_c \exp[-(E_c - E_F)/kT]$, and $n_t \approx N_t \exp[-(E_t - E_F)/kT]$.

### Decrease in photosensitivity

If there are no traps present, then

$$G = n/\tau = nS_r vN_r^* = n^2 S_r v \qquad (2.36)$$

where $N_r^*$ is the density of available empty (hole-occupied) recombination centers with cross section $S_r$. For every free electron there is a corresponding hole either in the valence band, $p$, or in $N_r^*$, so that $n = N_r^* + p$ and it is reasonable to assume $N_r^* \gg p$. Therefore from Eq. (2.36) the lifetime without traps is $\tau_1 = 1/nvS_r$.

With traps, $N_r^* = n + N_t \approx N_t$, where $N_t$ is the total density of trapped electrons, and the lifetime is $\tau_2 = 1/(N_t vS_r)$. It is evident that the lifetime has been reduced by the factor $n/N_t$.

## 2.5    Simple kinetics: intrinsic semiconductor

To get a feeling for the types of photoconductivity behavior that can be obtained even in a very simple system, we consider here briefly the case of intrinsic photoexcitation in an intrinsic semiconductor according to the diagram of Figure 2.4(a). Photoexcitation rate $G$ and thermal excitation rate $g$ are balanced by recombination across the bandgap with a rate of $R$.

In the dark

$$g = n_0 S v p_0 = n_0^2 S v \qquad (2.37)$$

since $n_0 = p_0$, with a carrier lifetime $\tau_d$ in the dark given by $\tau_d = 1/(n_0 S v)$.

In the light

$$G + g = (n_0 + \Delta n)Sv(p_0 + \Delta p) = (n_0 + \Delta n)^2 vS \qquad (2.38)$$

since $\Delta n = \Delta p$. We may now distinguish several representative cases.

*Insulator case*

In the insulator case, $g \ll G$, $n_0 \ll \Delta n$, and Eq. (2.38) becomes

$$G = (2n_0 \Delta n + \Delta n)^2 vS \approx \Delta n^2 vS. \tag{2.39}$$

Therefore

$$\Delta n = (G/vS)^{1/2} \tag{2.40}$$

$$\tau = 1/vS \, \Delta n = [1/(vSG)]^{1/2} \tag{2.41}$$

In this case $\Delta n \propto G^{1/2}$, a case of sublinear photoconductivity. Correspondingly the lifetime decreases with increasing photoexcitation rate, $\tau \propto G^{-1/2}$. Cases like this in which $G \propto \Delta n^2$ are often called cases of 'bimolecular' recombination.

The situation can be changed radically simply by the addition of traps to the insulator, as in Figure 2.4(b). Now capture of electrons by traps is included together with the recombination of electrons and holes,

$$G = (n_0 + \Delta n)vS_t(N_t - n_t) + (n_0 + \Delta n)vS(p_0 + \Delta p) \tag{2.42}$$

where $N_t$ is the total density of traps, and $n_t$ is the density of electron-occupied traps. For the insulator case, Eq. (2.42) becomes

$$G = \Delta n vS_t(N_t - n_t) + \Delta n \, \Delta p \, vS \tag{2.43}$$

with $\Delta n$ no longer equal to $\Delta p$. To simplify the interpretation, suppose that the traps are all filled after some period of photoexcitation so that $n_t = N_t$. Then $G = \Delta n \, \Delta p \, vS$, and $\Delta p = \Delta n + N_t \approx N_t$, so that $G = \Delta n \, vSN_t$. The presence of the traps has changed the behavior so that

Figure 2.4. Energy band diagrams for (a) an intrinsic semiconductor under intrinsic photoexcitation, and (b) a semiconductor with traps under intrinsic photoexcitation.

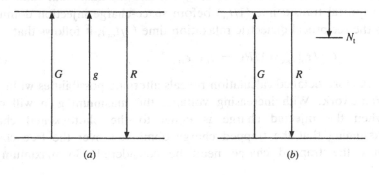

now $\Delta n \propto G$ and the lifetime is constant, $\tau = 1/\upsilon S N_t$, as shown at the end of the previous section.

### Semiconductor case

In the semiconductor case, we assume that $\Delta n \ll n_0$ and Eq. (2.38) becomes

$$\Delta n = G/2n_0\upsilon S \tag{2.44}$$

Thus $\Delta n \propto G$ and the lifetime is constant,

$$\tau = 1/2\upsilon S n_0 \tag{2.45}$$

In this case the lifetime is controlled by the dark properties of the semiconductor and indeed $\tau = \tau_d/2$.

## 2.6    Maximum gain

As stated in Section 2.2, the achievement of photoconductivity gains in excess of unity requires the presence of ohmic contacts. The magnitude of the gain can be increased by increasing the applied voltage, thus decreasing the transit time of carriers. But, on the other hand, the application of a high voltage to an ohmic contact will eventually lead to space-charge-limited current injection, as described in Section 1.4. It therefore follows that there is a maximum gain that can be achieved.

A simple approach to deriving a value for this maximum gain is the following. Consider the product of gain $G^*$ and $1/\tau_0$, where $\tau_0$ is the response time of the photoconductor, so that $G^*/\tau_0$ is the gain-bandwidth product, and inquire as to the maximum value of this gain-bandwidth product. Since $G^* = \tau/t_{tr}$,

$$G^*/\tau_0|_{max} = [\tau/\tau_0]_{max}[1/t_{tr}]_{max} \tag{2.46}$$

Since the maximum value of $\tau/\tau_0 = 1$, and the maximum value of the reciprocal transit time $(1/t_{tr})$ before space-charge injection dominates is the reciprocal dielectric relaxation time $(1/t_{dr})$, it follows that

$$G^*/\tau_0|_{max} = 1/RC = \sigma/\epsilon_r\epsilon_0 \tag{2.47}$$

A more detailed calculation reveals alternate possibilities within this framework. With increasing voltage, the maximum gain will occur when the injected charge is equal to the photoexcited charge. Assuming that the trapped charge dominates over the free charge, only the trapped charge need be considered. So maximum $G^*$

corresponds to

$$qn_t^V dA = qn_t^L dA \tag{2.48}$$

where $n_t^V$ is the trapped charge due to field injection, and $n_t^L$ is the trapped charge due to photoexcitation. The term on the left of Eq. (2.48) is just equal to the product of the applied voltage $\phi_{max}$ for maximum gain and the capacitance: $\phi_{max}C$. The term on the right can be written in terms of the response time $\tau_0 = \tau(n_t^L/n)$ as $qn(\tau_0/\tau)dA$. If, therefore, it is assumed that $n_t^V = n_t^L$, then $G_{max}^* = \tau\mu\phi_{max}/d^2$, and

$$G_{max}^*/\tau_0 = \sigma A/dC = 1/RC \tag{2.49}$$

since $\phi_{max} = qn\tau_0 dA/\tau C$, in agreement with Eq. (2.47).

If, on the other hand, it were possible that $n_t^V \neq n_t^L$, then

$$G_{max}^* = \tau\mu\phi_{max}/d^2 = \tau(\mu qn_t^V A/dC) = (n\tau_0/n_t^L)(\mu qn_t^V A/dC), \text{ and}$$

$$G_{max}^*/\tau_0 = (\sigma\tau_0 A/dC)(n_t^V/n_t^L) = (1/RC)(n_t^V/n_t^L) \tag{2.50}$$

and it is at least formally possible that the maximum gain may exceed $1/RC$ by the ratio $n_t^V/n_t^L = M$. Values of $M$ larger than unity require the existence of conditions such that electrically injected carriers will be trapped but photoexcited carriers will not. In a homogeneous material, recombination centers lying just above the Fermi level could meet these requirements; in an inhomogeneous system, surface traps may capture the injected charge, while photocurrent flows primarily through volume regions with a much smaller trap density. Experimental values of $M$ as high as 500 have been reported.[1]

## 2.7 Types of photoconducting systems

A number of different types of photoconducting systems are commonly encountered. Understanding them allows us to appreciate a broad range of photoelectronic phenomena in semiconductors.

### Homogeneous material

Five different possibilities for photoconducting behavior in homogeneous material are pictured in Figure 2.5, depending on the behavior of the contacts, and the importance of photoexcited carrier capture.

Figure 2.5(a) shows the case of a material with ohmic contacts for both electrons and holes, and without imperfection states. Photoexcitation produces free electrons and holes that move through the material, being replaced when they leave the material at a contact, until they recombine with one another. The gain is simply the sum of the electron gain and the hole gain,

$$G^* = \tau_n/t_{trn} + \tau_p/t_{trp} = (\tau_n\mu_n + \tau_p\mu_p)\phi/d^2 \tag{2.51}$$

Figure 2.5($b$) is similar except that the contact for electrons is ohmic but the contact for holes is not. The gain is once again the sum of the electron gain and the hole gain. But this time, the effective hole lifetime is the hole transit time, because once the hole leaves the material, it is not replaced. Also the effective electron lifetime is also the hole transit time, because once the hole leaves the material, there is no need to replace the electron from the contact to maintain charge neutrality. The result is that

$$G^* = t_{trp}/t_{trn} + 1 = (\mu_n + \mu_p)/\mu_p \tag{2.52}$$

Figure 2.5. Five basic types of homogeneous photoconductor.

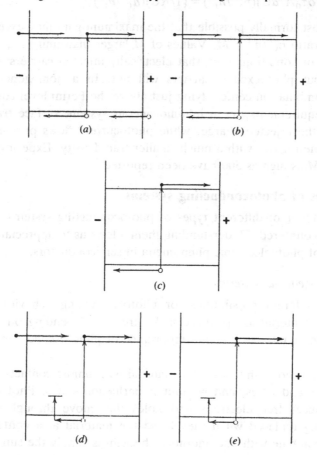

Since usually $\mu_n > \mu_p$, this situation can also lead to gains greater than unity.

Figure 2.5(c) shows the case where neither contact is ohmic. The lifetime for both electrons and holes is simply their corresponding transit times, and $G^* = 1$.

In Figure 2.5(d) we introduce the possibility of capture of a hole at an imperfection in the material, along with the presence of a non-ohmic contact for holes. The photoexcited hole is captured at the imperfection, constituting a positive charge in the material. When a photoexcited electron leaves the material, it is replaced at the negative contact until recombination occurs between the free electron and the captured hole. The gain is effectively due only to electrons and is given by the simple expression $G^* = \tau_n \mu_n \phi / d^2$. This is the 'normal' situation in a real homogeneous photoconductor showing gain greater than unity.

Finally Figure 2.5(e) pictures the case where the hole is captured at an imperfection and both contacts are non-ohmic. In this case the material polarizes, the trapped charge cancels the applied field, and no steady photoconductivity can flow.

Historically it is of some interest to note that the cases of $G^* = 1$ were for many years considered to be the authentic effect because of the obvious correlation with quantum views of light, and such currents were called 'primary photocurrents.' Occasionally situations were encountered in which $G^* > 1$, but these were regarded with suspicion and were labeled 'secondary photocurrents.'

### p–n junctions

A typical p–n junction is shown in Figure 2.6. Electrons excited in the p-type material by photoexcitation, or holes excited in the n-type material by photoexcitation, can diffuse to the junction if their point of origin is not much larger than a diffusion length from the junction interface. Once an electron or a hole has been collected, however, it cannot be replaced and the maximum value of $G^* = 1$, one net electronic charge being collected per photon absorbed.

In many ways the p–n junction is similar to the case of Figure 2.5(c). The p-type semiconductor plays the role of a non-ohmic contact for electrons to the depletion region, and the n-type semiconductor plays the role of a non-ohmic contact for holes to the depletion region.

Since a p–n junction can be viewed for some purposes as equivalent to two Schottky barriers back-to-back, it is not surprising that a

Figure 2.6. Energy band diagram for a p–n junction.

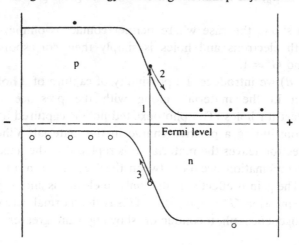

Figure 2.7. Energy band diagrams for metal–semiconductor Schottky barriers on (*a*) n-type and (*b*) p-type material.

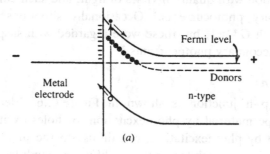

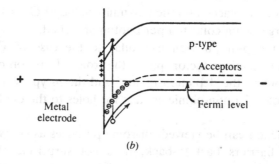

Schottky barrier between a metal and a semiconductor falls into the same class of photoconductors. Figure 2.7 shows energy band diagrams for Schottky barriers to n- and p-type materials. If a thin metal film is used so that most of the light can penetrate into the semiconductor, a Schottky barrier can be used as a photoconducting device with maximum $G^* = 1$.

### n–p–n junctions

Gains larger than unity can be obtained with junction devices provided that arrangements are made to provide the possibility of carrier replenishment to maintain charge neutrality in a critical region. Figure 2.8 shows two ways of accomplishing this with either an n–p–n or a p–n–p structure. We discuss the n–p–n structure since material parameters usually favor this form.

Qualitatively the behavior is as follows for Figure 2.8(a) with a corresponding description appropriate for Figure 2.8(b). Electron–hole pairs formed (1) by photoexcitation within a diffusion length of

Figure 2.8. Energy band diagrams for (a) an n–p–n junction and (b) a p–n–p junction.

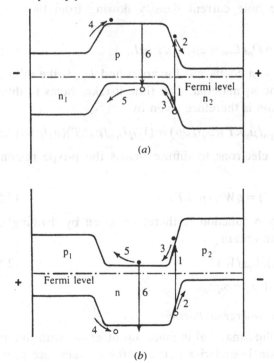

(a)

(b)

the $p-n_2$ junction result in the collection of the electron (2) through $n_2$ and the presence of the hole (3) in the p-type potential well formed by the junction structure. As long as the hole is in the p-type region, it constitutes a positive charge that forward biases the $n_1-p$ junction and injects an electron (4) to replace the one that was originally produced by photoexcitation. This process continues until either the hole recombines with a passing electron (6), or the hole diffuses out of the p-material (5) into the $n_1$ material, usually the more likely process. The gain of the device can be considered in the same formalism as that used for the homogeneous material in Figure 2.5($d$), to which the n–p–n junction is conceptually quite similar. The gain is equal to the lifetime of the electrons passing through the p-type region, which, in turn, is equal to the time it takes for the hole to diffuse out of the p-type region. The transit time is the transit time for electrons to diffuse across the p-type region.

If $p$ holes are injected into the p-type region, this reduces the $n_1-p$ junction by $\Delta E$. This, in turn, increases the electron flow from $n_1$ into the p-type region by $\Delta n = n_1 \exp(\Delta E/kT) = \Delta p$, and increases the hole flow from the p-type region to $n_1$ by $\partial p = p \exp(\Delta E/kT)$. If the p-type region has a width $W$, the total charge per unit area in the p region is $W\Delta p$. The hole current density flowing from the p-type region is

$$J_p = \partial p\, v = \partial p\, D_p/L_{pn} = \partial p (\mu_p kT/qL_{pn}) \qquad (2.53)$$

where $D_p$ is the diffusion coefficient for holes and $L_{pn}$ is the diffusion length for holes in the $n_1$ material. The time it takes holes to diffuse out of the p-type region is therefore given by

$$t_p = W\Delta p/J_p = (WqL_{pn}/\mu_p kT)(\Delta p/\partial p) = (WqL_{pn}/\mu_p kT)(n_1/p) \quad (2.54)$$

The transit time for electrons to diffuse across the p-type region is given by

$$t_n = W/(D_n/W) = qW^2/(\mu_n kT) \qquad (2.55)$$

The gain of the n–p–n junction is therefore given by dividing Eq. (2.54) by Eq. (2.55) to obtain

$$G^* = (\sigma_{n1}/\sigma_p)(L_{pn}/W) \qquad (2.56)$$

where $\sigma_{n_1} = n_1 q\mu_n$ and $\sigma_p = pq\mu_p$.

### Polycrystalline intergrain barriers

A polycrystalline material is made up of grains with different orientations. The grain boundaries between these grains are characterized by interface states leading to a potential barrier between the grains. Photoexcitation can affect transport across these grain boundaries by (*a*) increasing the density of free carriers throughout the material, (*b*) decreasing the intergrain barrier height by changing the charge in the intergrain states, and (*c*) increasing the possibility for tunneling through the intergrain barriers by decreasing the depletion layer widths in the adjacent grains.

Three traditional ways of looking at barriers and photoconductivity phenomena are shown in Figure 2.9. In Figure 2.9(*a*) the only effect of photoexcitation is to increase the density of free carriers in the intergrain regions; this is treated by a standard photoconductivity model and is not really a barrier problem. In Figure 2.9(*b*) the effect of photoexcitation is primarily to increase the density of the free

Figure 2.9. Different ways of looking at barriers and photoconductivity-related phenomena. (*a*) Photoexcitation increases only the density of free carriers in the barrier region. (*b*) Photoexcitation increases the density of free carriers in the grains, and hence increases the density of carriers in the barrier regions as well. (*c*) Photoexcitation does not change the carrier density in the grains, but decreases the barrier height by depositing suitable trapped charge in or near the barrier region.

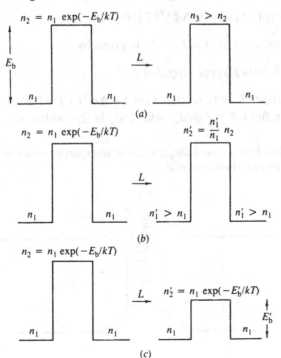

carriers in the grains, and hence as a result to increase the density of the free carriers in the intergrain regions; such a model has been called a 'numbers' model. In Figure 2.9(c) the effect of photoexcitation is primarily to decrease the barrier height; such a model has been called a 'mobility modulation' model. Much attention has been devoted to particular polycrystalline materials to determine whether a numbers model or a mobility modulation model was the appropriate one. Although examples do exist where one or the other dominates, in many cases not surprisingly both phenomena occur simultaneously.

Figure 2.10 shows an energy band diagram of an intergrain barrier in an n-type polycrystalline material. Without an applied voltage electrons flow in both directions across the barrier and produce no net current. If a voltage difference $\Delta\phi$ is applied to the grain boundary such that the right side of the material in Figure 2.10 is positive, then the effective barrier height on the right is increased from $q\phi_b$ to $(q\phi_b + q\,\Delta\phi)$ and the effective barrier height on the left is decreased from $q\phi_b$ to $(q\phi_b - q\,\Delta\phi)$. Thus the current flowing to the right can now be written as

$$J_\rightarrow \approx \tfrac{1}{2}n\,\exp[-(q\phi_b - q\,\Delta\phi)/kT]qv \qquad (2.57)$$

and the current flowing to the left is

$$J_\leftarrow \approx \tfrac{1}{2}n\,\exp[-(q\phi_b + q\,\Delta\phi)/kT]qv \qquad (2.58)$$

Therefore the total current, $J = J_\rightarrow - J_\leftarrow$ is given by

$$J = (nq^2\Delta\phi v/kT)\,\exp(-q\phi_b/kT) \qquad (2.59)$$

provided that $q\,\Delta\phi/kT \ll 1$, so that $\exp(\pm q\,\Delta\phi/kT) \approx 1 \pm q\,\Delta\phi/kT$. Since the electric field $\mathscr{E} = \Delta\phi/d_g$, where $d_g$ is the width of a grain,

Figure 2.10. Energy band diagram for an interparticle grain barrier in an n-type polycrystalline material.

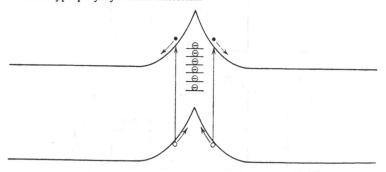

and $J = \sigma \mathscr{E}$, we may solve for the conductivity,

$$\sigma = (nq^2 d_g v/kT) \exp(-q\phi_b/kT) \tag{2.60}$$

If we choose to express $\sigma = nq\mu^*$, where $\mu^*$ represents the grain-barrier controlled mobility for transport through the material,

$$\mu^* = (qv d_g/kT) \exp(-q\phi_b/kT) \tag{2.61}$$

Therefore $\mu^* = \mu_0^* \exp(-q\phi_b/kT)$ and measurements of the effective mobility in the polycrystalline material should be thermally activated with an activation energy of $q\phi_b$.

The effects of illumination can be represented by

$$\Delta\sigma = q\mu^* \Delta n + nq \Delta\mu^* \tag{2.62}$$

with

$$\Delta\mu^* = -[q^2 v d_g/(kT)^2] \Delta\phi_b \exp(-q\phi_b/kT) \tag{2.63}$$

If we define $B = (\Delta\mu^*/\mu^*)/(\Delta n/n_0) = (n_0 q/kT)(\Delta\phi_b/\Delta n)$, then

$$\Delta\sigma = q\mu^*(1 + B) \Delta n \tag{2.64}$$

## 2.8 Mathematical models for photoconductivity

The basic equations needed for a complete description of photoconductivity effects can be written down fairly readily in principle, but their detailed solution is often much more complicated in the presence of a number of different types of imperfections. In this section we take a look at the general mathematical background and some of the typical simplifications, together with another approach that enables one to describe many phenomena conveniently without a detailed mathematical solution.

### Complete mathematical model

A complete mathematical model consists of continuity equations for electrons and holes,

$$\partial n/\partial t = G - \sum_i [n\beta_i(N_i - n_i)]$$

$$+ \sum_i \{n_i N_c \beta_i \exp[-(E_c - E_i)/kT]\} + \nabla \cdot \mathbf{J}_n/q \tag{2.65}$$

where the second term on the right represents the capture of free electrons by imperfections, and the third term on the right represents the thermal excitation of electrons occupying imperfections to the conduction band. $N_i$ is the total density of the $i$th imperfection, $n_i$ is the density of electron-occupied $i$th imperfections, $\beta_i$ is the capture coefficient of the $i$th imperfection for free electrons, and $(E_c - E_i)$ is the ionization energy of the $i$th imperfections.

$$\partial p / \partial t = G - \sum_j (p\beta_j n_j)$$

$$+ \sum_j \{(N_j - n_j)N_v\beta_j \exp[-(E_j - E_v)/kT]\} + \nabla \cdot \mathbf{J}_p/q$$

(2.66)

where the second term on the right represents the capture of free holes by imperfections, and the third term on the right represents the thermal excitation of holes occupying imperfections to the valence band. $N_j$ is the total density of the $j$th imperfection, $n_j$ is the density of electron-occupied $j$th imperfections, $\beta_j$ is the capture coefficient of the $j$th imperfection for free holes, and $(E_j - E_v)$ is the ionization energy of the $j$th imperfections.

Electron and hole current densities can be defined as in Eqs. (1.6)–(1.12):

$$\mathbf{J}_n = nq\mu_n \mathscr{E} + qD_n \nabla n \qquad (2.67)$$

$$\mathbf{J}_p = pq\mu_p \mathscr{E} - qD_p \nabla p \qquad (2.68)$$

Finally there is Poisson's Equation:

$$\nabla \cdot \mathscr{E} = Q/\epsilon_r\epsilon_0 \qquad (2.69)$$

where $Q$ is the net space charge in the material.

Combining these equations gives for electrons, for example,

$$\partial n / \partial t = G - \sum_i [n\beta_i(N_i - n_i)] + \sum_i \{n_i N_c \beta_i \exp[-(E_c - E_i)/kT]\}$$

$$+ D_n \nabla^2 n + \mu_n \mathscr{E} \cdot \nabla n + \mu_n n Q/\epsilon_r\epsilon_0 \qquad (2.70)$$

The following simplifying assumptions are frequently adopted:

(1)   $Q = 0$   Neglect space charge; assume charge neutrality.
(2)   $\nabla n = \nabla p$   Neglect diffusion in the bulk of the material.
(3)   Replace

$$\left(-\sum_i [n\beta_i(N_i - n_i)] + \sum_i \{n_i N_c \beta_i \exp[-(E_c - E_i)/kT]\}\right)$$

by $-n/\tau_n$, in terms of an appropriately defined liftime $\tau_n$.

If all three of these simplifying assumptions are made, the problem reduces in steady state to the very simple result: $G = n/\tau_n$ and

$G = p/\tau_p$, and the burden falls upon the definition of the appropriate lifetimes corresponding to the imperfections present.

### Steady state Fermi levels

The thermal equilibrium Fermi level is a useful construct because knowledge of its location permits the description of the occupancy of all other levels in the semiconductor. Even under illumination, however, the concept of a Fermi level – now a steady state Fermi level – can be usefully retained.

Figure 2.11(a) shows the normal energy band diagram in thermal equilibrium. We can define the Fermi level position either as

$$E_c - E_F = kT \ln(N_c/n_0) \tag{2.71}$$

or as

$$E_F - E_v = kT \ln(N_v/p_0) \tag{2.72}$$

It follows that the sum of Eqs. (2.71) and (2.72) is given by

$$(E_c - E_F) + (E_F - E_v) = kT \ln(N_c N_v/n_0 p_0) = E_c - E_v = E_G \tag{2.73}$$

where $E_G$ is the bandgap of the semiconductor.

Under illumination we define the steady state Fermi levels by the same form as Eqs. (2.71) and (2.72), except that now there must be a separate electron steady state Fermi level $E_{Fn}$ and hole steady state Fermi level $E_{Fp}$, since $n = n_0 + \Delta n$, and $p = p_0 + \Delta p$, as shown in Figure 2.11(b). The steady state Fermi levels are defined as

$$E_c - E_{Fn} = kT \ln(N_c/n) \tag{2.74}$$

Figure 2.11. Energy band diagrams showing (a) the thermal equilibrium Fermi level, and (b) the steady state electron and hole Fermi levels.

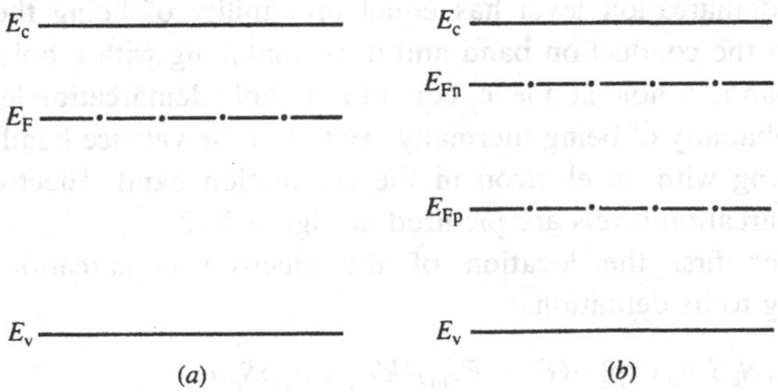

and as

$$E_{Fp} - E_v = kT \ln(N_v/p) \tag{2.75}$$

The difference between the two steady state Fermi levels is given by

$$E_{Fn} - E_{Fp} = kT \ln(np/n_0 p_0) \tag{2.76}$$

and their sum is given by

$$(E_c - E_{Fn}) + (E_{Fp} - E_v) = kT \ln(N_c N_v/np)$$

$$= E_G - kT \ln(np/n_0 p_0) \tag{2.77}$$

where $n_0 p_0 = n_{int}^2$, with $n_{int}$ the intrinsic carrier density in the semiconductor.

The difference between the electron steady state Fermi level $E_{Fn}$ and the electron thermal equilibrium Fermi level $E_F$ is given by

$$E_F - E_{Fn} = kT \ln[n_0/(n_0 + \Delta n)] \tag{2.78}$$

The steady state Fermi levels describe the occupancy of those levels that are still in effective thermal equilibrium with the bands, i.e., with those imperfections whose occupancy is determined by thermal exchange with the bands rather than by recombination. An alternate way of expressing this is to say that the steady state Fermi level continues to describe the occupancy of traps during illumination, but not of recombination centers.

### Demarcation levels

To describe this demarcation between levels whose occupancy is controlled by thermal exchange with the band, and those levels whose occupancy is controlled by recombination, we may define a 'demarcation level.' Clearly we need one demarcation level for electrons and another for holes. An electron at a level equal to the electron demarcation level has equal probability of being thermally excited to the conduction band and of recombining with a hole in the valence band. A hole at a level equal to the hole demarcation level has equal probability of being thermally excited to the valence band and of recombining with an electron in the conduction band. Electron and hole demarcation levels are pictured in Figure 2.12.

Consider first the location of the electron demarcation level. According to its definition,

$$n_1 N_c S_n v_n \exp[-(E_c - E_{Dn})/kT] = n_1 p S_p v_p \tag{2.79}$$

where $n_I$ is the density of electron-occupied imperfection levels with total density $N_I$, and with capture cross section $S_n$ for electrons when empty, and $S_p$ for holes when occupied. Two possible substitutions may be made: either we may substitute $N_c = n \exp[(E_c - E_{Fn})/kT]$ or we may substitute $p = N_v \exp[-(E_{Fp} - E_v)/kT]$. In the first case, Eq. (2.79) becomes

$$E_c - E_{Dn} = (E_c - E_{Fn}) - kT \ln[(S_p v_p/S_n v_n)(p/n)] \qquad (2.80)$$

In the second case it becomes

$$E_c - E_{Dn} = (E_{Fp} - E_v) + kT \ln[(m_e*/m_h*)(S_n/S_p)] \qquad (2.81)$$

where $m_e*$ and $m_h*$ are the effective masses of electrons and holes respectively.

The equation for holes, corresponding to Eq. (2.79), is

$$(N_I - n_I)N_v S_p v_p \exp[-(E_{Dp} - E_v)/kT] = n(N_I - n_I)S_n v_n \qquad (2.82)$$

Substitution can be made either for $N_v$ on the left, or for $n$ on the right. The results are

$$(E_{Dp} - E_v) = (E_{Fp} - E_v) - kT \ln [(S_n v_n/S_p v_p)(n/p)] \qquad (2.83)$$

$$(E_{Dp} - E_v) = (E_c - E_{Fn}) + kT \ln[(m_h*/m_e*)(S_p/S_n)] \qquad (2.84)$$

The utility of these expressions may be illustrated by the following example. Suppose that a measurable phenomenon occurs when a particular set of levels changes from recombination centers to hole traps in a material in which the conductivity is controlled by electrons. A measurement of electron density at the critical point then pinpoints the location of the demarcation level according to Eq. (2.84) through a constant that depends directly on the cross section ratio.

Figure 2.12. Energy band diagrams showing (*a*) the electron demarcation level and (*b*) the hole demarcation level.

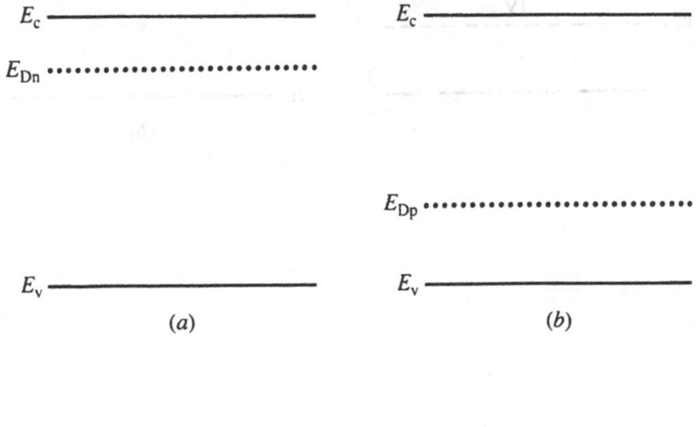

|       (a)       |       (b)       |

Figure 2.13($a$) shows a typical situation in an insulator ($\Delta n \gg n_0$). Imperfection levels in bandgap region I are effectively in thermal equilibrium with the conduction band, and have their occupancy described by the steady state electron Fermi level; they are electron traps. Imperfection levels in bandgap region V are effectively in thermal equilibrium with the valence band, and have their occupancy described by the steady state hole Fermi level; they are hole traps. Imperfection levels in bandgap regions III and IV lie between the demarcation levels, have their occupancy determined by kinetics and not by the location of the steady state Fermi levels; they are recombination centers. Imperfection levels in II are in effective thermal equilibrium with the conduction band, but since they do have their occupancy described by the steady state electron Fermi level, they are mostly occupied, and hence are usually counted with the recombination centers.

In an n-type semiconductor ($\Delta n \ll n_0$) the situation is typically described by the diagram in Figure 2.13($b$). Electron traps are found in I, and hole traps in V. Region III is missing, and recombination levels are found in the remaining central region of the bandgap in which II and IV are now equivalent.

Figure 2.13. Typical energy band diagrams showing steady state Fermi levels and demarcation levels for ($a$) an insulator, and ($b$) a semiconductor.

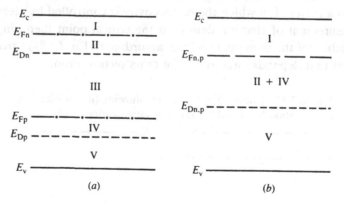

# 3

# One-center recombination models

The next step beyond the intrinsic recombination model discussed in Section 2.5 is a model in which recombination is dominated by a single type of imperfection level. In this chapter we consider several variations on this theme.

## 3.1 The simplest one-center model

Consider the model pictured in Figure 3.1. The only transitions considered are optical excitation from the valence band to the conduction band, and electron and hole capture at an imperfection present with total density $N_I$. No thermal exchange with the bands is included. No other levels such as traps are considered. The material is assumed to be an insulator so that $\Delta n \gg n_0$. We also assume that the location of the Fermi level is fixed above the imperfection level by interaction between the imperfections and some appropriate density $N_D$ of very shallow donors whose occupancy is unaffected by photoexcitation.

In this case we have three continuity equations:

$$dn/dt = G - n\beta_n(N_I - n_I) \tag{3.1}$$

$$dn_I/dt = n\beta_n(N_I - n_I) - n_I p\beta_p \tag{3.2}$$

$$dp/dt = G - n_I p\beta_p \tag{3.3}$$

It is evident that only two of these are independent.

Another equation that we can make use of is one that can be called 'particle conservation.' Every electron found above the Fermi level under photoexcitation, must have a corresponding hole below the Fermi level. If it is assumed that all of the imperfection levels are occupied in the dark, then

$$n = p + (N_I - n_I) \tag{3.4}$$

This condition is equivalent to charge neutrality and is often more convenient to use. To see this, consider the dark and light charge neutrality conditions:

$$\text{Dark:} \quad n_0 + N_I = N_D + p_0 \tag{3.5a}$$

$$\text{Light:} \quad n_0 + \Delta n + n_I = N_D + p_0 + \Delta p \tag{3.5b}$$

Subtracting Eq. (3.5a) from Eq. (3.5b) gives

$$\Delta n = \Delta p + (N_I - n_I) \tag{3.5c}$$

which in the insulator approximation of $\Delta n$, $\Delta p \gg n_0$, $p_0$ is identical to Eq. (3.4).

### *n-equation in steady state*

We have the following equations that we can use to determine the dependence of $n$ on $G$, from Eqs. (3.1), (3.3) and (3.4):

$$G = n\beta_n(N_I - n_I) \tag{3.6}$$

$$G = n_I p \beta_p \tag{3.7}$$

$$n = p + (N_I - n_I) \tag{3.8}$$

Substitution of Eq. (3.8) in Eq. (3.7) gives

$$G = n_I(n - N_I + n_I)\beta_p \tag{3.9}$$

Solve for $n_I$ from Eq. (3.6)

$$n_I = N_I - G/n\beta_n \tag{3.10}$$

Figure 3.1. Energy band diagram for simple one-level recombination model. Recombination centers with density $N_I$ are characterized by an electron capture coefficient $\beta_n$ and a hole capture coefficient $\beta_p$.

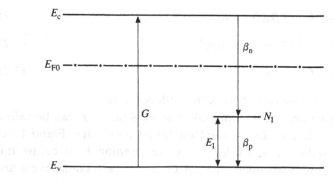

Substitute Eq. (3.10) into Eq. (3.9) to obtain

$$G = (N_I - G/n\beta_n)(n - G/n\beta_n)\beta_p \tag{3.11}$$

This is then the basic relationship between $G$ and $n$. As an equation in $n$ it is a cubic; as an equation in $G$ it is a quadratic.

### p-equation in steady state

Exactly the same procedure can be followed to obtain the $p$ vs $G$ dependence. Substitute Eq. (3.8) into Eq. (3.6) and solve for $(N_I - n_1)$ from Eq. (3.7). The final result is

$$G = (p + N_I - G/p\beta_p)(N_I - G/p\beta_p)\beta_n \tag{3.12}$$

### Qualitative evaluation

Before continuing with the quantitative calculation, let us look at the problem and see what kind of results are expected simply from the basic phenomena involved.

For small values of $n$ or low-intensity photoexcitation, there is a hole in a recombination center for every electron in the conduction band: $n \approx (N_I - n_1)$. Therefore $G = n^2\beta_n$ and $n = (G/\beta_n)^{1/2}$. We predict therefore that for small $n$, $n \propto G^{1/2}$.

For small $n$, almost all of the recombination centers are filled, so that $n_1 = N_I$. Therefore for holes, $G = pN_I\beta_p$ and $p = G/\beta_pN_I$. We predict therefore for small $n$, $p \propto G$.

For large values of $n$ or high-intensity photoexcitation, suppose that $\beta_n \ll \beta_p$, so that almost all of the recombination centers are empty: $N_I - n_1 \approx N_I$. Then electron capture will dominate the kinetics and we expect $n = p = G/\beta_nN_I$. Therefore for large values of $n$ under these conditions we predict that $n = p \propto G$.

### Specific solutions: n-equation

Eq. (3.11) becomes

$$G^2 - GnN_I\beta_n^2[1/\beta_n + (n/N_I)(1/\beta_n + 1/\beta_p)] + n^3\beta_n^2N_I = 0 \tag{3.13}$$

For small $n$, such that $(n/N_I)(1/\beta_n + 1/\beta_p) \ll 1/\beta_n$, the solution for the quadratic in $G$ can be written as

$$G = (nN_I\beta_n/2)[1 \pm (1 - 2n/N_I)] \tag{3.14}$$

using the approximation that $(1 - x)^{1/2} \approx (1 - x/2)$ for small $x$. The solution corresponding to the positive sign does not correspond to the physical situation (giving $G = nN_I\beta_n$ but the levels are all filled, not empty), and therefore we choose the solution for the negative sign: $G = n^2\beta_n$ or $n = (G/\beta_n)^{1/2}$.

For large $n$, when $(n/N_I)(1/\beta_n + 1/\beta_p) \gg 1/\beta_n$, the solution is

$$G = [n^2\beta_n^2(1/\beta_n + 1/\beta_p)/2](1 \pm \{1 - 2N_I/[n\beta_n^2(1/\beta_n + 1/\beta_p)^2]\}) \quad (3.15)$$

using the same approximation as in Eq. (3.14). Again the positive sign does not correspond to the physical situation (giving $G \propto n^2$ with no physical basis), and the negative sign gives

$$G = nN_I/(1/\beta_n + 1/\beta_p) \quad (3.16)$$

so that $n = (G/N_I)(1/\beta_n + 1/\beta_p)$.

### Specific solutions: p-equation

Eq. (3.12) expands to give

$$G^2 - GpN_I\beta_p^2[2/\beta_p + (p/N_I)(1/\beta_n + 1/\beta_p)] + N_Ip^2\beta_p^2(p + N_I) = 0 \quad (3.17)$$

For small values of $p$, $(p/N_I)(1/\beta_n + 1/\beta_p) \ll 2/\beta_p$, and the solution becomes $G = pN_I\beta_p$ or $p = G/\beta_pN_I$.

For large values of $p$, $(p/N_I)(1/\beta_n + 1/\beta_p) \gg 2/\beta_p$, and Eq. (3.17) becomes

$$G = [p^2\beta_p^2(1/\beta_n + 1/\beta_p)/2](1 \pm \{1 - 2N_I/[p\beta_p^2(1/\beta_n + 1/\beta_p)^2]\}) \quad (3.18)$$

Again the choice of the negative sign is the one with physical significance, giving $G = pN_I/(1/\beta_n + 1/\beta_p)$ and hence $p = (G/N_I) \times (1/\beta_n + 1/\beta_p)$. So we see that in every case our qualitative expectations are borne out by the detailed solution of the problem.

In the above calculations we did not specifically use the equation for continuity of $n_I$, Eq. (3.2). In steady state,

$$n_I/N_I = \beta_n n/(\beta_n n + \beta_p p) = 1/(1 + \beta_p p/\beta_n n) \quad (3.19)$$

As an illustration take the values of $n = (G/\beta_n)^{1/2}$ and $p = G/\beta_pN_I$ from the case of small $n$. Then Eq. (3.19) becomes

$$n_I = N_I/(1 + n/N_I) = N_I(1 - n/N_I) = N_I - n \quad (3.20)$$

as expected. As another illustration take the values for large $n$, $n = p = (G/N_I)(1/\beta_n + 1/\beta_p)$. Then

$$n_I/N_I = 1/(1 + \beta_p/\beta_n) \quad (3.21)$$

and we have either that $n_I/N_I = \beta_n/\beta_p$, if $\beta_n \ll \beta_p$, or $n_I/N_I = 1$, if $\beta_n \gg \beta_p$. A plot of the variation of $n$ and $p$ with $G$ for typical values of the constants is given in Figure 3.2.

### A simple computer approach

Start with the three basic equations for particle conservation, continuity of $n_I$, and continuity of $p$:

$$n = p + (N_I - n_I) \tag{3.22}$$

$$n_I/N_I = \beta_n n/(\beta_n n + \beta_p p) \tag{3.23}$$

$$G = p\beta_p n_I + \beta_{cv} np \tag{3.24}$$

where the second term on the right of Eq. (3.24) generalizes the problem to include the case of direction recombination between free electrons and free holes with a capture coefficient of $\beta_{cv}$.

Substitute Eq. (3.23) into Eq. (3.22) and use the computer to calculate $p(n)$. Then substitute this result in Eq. (3.24) to obtain $G(n)$, and hence $n(G)$ and $p(G)$.

In the next step of generality, we consider the possibility for a number of different recombination centers, but continue to neglect thermal excitation processes. Eqs. (3.22)–(3.24) are readily general-

Figure 3.2. Variation of electron density (open circles), hole density (solid circles), and density of electron-occupied imperfections (squares), for the simple one-center recombination model of Figure 3.1. Calculated for $N_I = 10^{16}\,\text{cm}^{-3}$, $S_n = 10^{-16}\,\text{cm}^2$, and $S_p = 10^{-12}\,\text{cm}^2$.

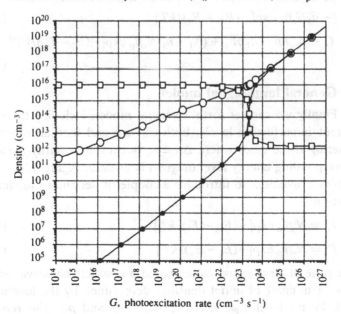

*G*, photoexcitation rate (cm$^{-3}$ s$^{-1}$)

ized to:

$$n + \sum_i n_i = p + \sum_j (N_j - n_j) \tag{3.25}$$

where the sum on the left is over all levels with energy $(E_i - E_v) >$ $(E_F - E_v)$, it being assumed that these are essentially empty in the dark, and where the sum on the right is over all levels with energy $(E_j - E_v) < (E_F - E_v)$, it being assumed that these are essentially electron-occupied in the dark.

$$n_i/N_i = \beta_{in}n/(\beta_{in}n + \beta_{ip}p) \tag{3.26}$$

$$G = p \sum_i \beta_{pi}n_i + \beta_{cv}np \tag{3.27}$$

## 3.2    Insulator model including thermal processes

The insulator model of Figure 3.1 can be generalized to include not only capture transitions between the bands and the imperfection level, but also thermal excitation transitions. In this case the three continuity equations for electrons, electron-occupied imperfections, and holes become:

$$dn/dt = G - n\beta_n(N_I - n_I) - \beta_{cv}np + n_I N_c \beta_n \exp[-(E_G - E_I)/kT] \tag{3.28}$$

$$dn_i/dt = n\beta_n(N_I - n_I) - n_I p\beta_p + (N_I - n_I)N_v\beta_p \exp[-(E_I - E_v)/kT]$$
$$\qquad - n_I N_c \beta_n \exp[-(E_G - E_I)/kT] \tag{3.29}$$

$$dp/dt = G - \beta_{cv}np - n_I p\beta_p + (N_I - n_I)N_v\beta_p \exp[-(E_I - E_v)/kT] \tag{3.30}$$

## 3.3    General large-signal model

Finally we can set forth a general model, including a number of different imperfection levels, both capture and thermal processes, and covering those cases where $\Delta n \geq n_0$ but not necessarily $\Delta n \gg n_0$. The corresponding energy level diagram is given in Figure 3.3.

It proves convenient to introduce a couple of definitions to simplify the notation:

$$P_{ic} = N_c\beta_{in} \exp[-(E_G - E_i)/kT] \tag{3.31}$$

$$P_{iv} = N_v\beta_{ip} \exp[-(E_i - E_v)/kT] \tag{3.32}$$

In the general large-signal model based on particle conservation, we start with a number of initial densities determined by the location of the dark Fermi level $E_{F0}$: $n_{i0}$, $(N_i - n_{i0})$, $n_0$ and $p_0$. The required equations are the following:

$$(n - n_0) + \sum_i (n_i - n_{i0}) = (p - p_0) + \sum_j [(N_j - n_j) - (N_j - n_{j0})]$$

$$(3.33)$$

The sum on the left is taken over all levels with $(E_i - E_v) > (E_{F0} - E_v)$, and the sum on the right is taken over all levels with $(E_j - E_v) < (E_{F0} - E_v)$. The increase in the electron occupancy of levels above the Fermi level upon illumination is equal to the increase in hole occupancy of levels below the Fermi level upon illumination. The dark values are given by

$$n_{i0}/N_i = 1/\{g \exp[(E_i - E_{F0})/kT] + 1\} \tag{3.34}$$

where $g$ is related to the imperfection degeneracy and is, in simple cases, equal to $\frac{1}{2}$ for electron-occupied donors, or 2 for electron-occupied acceptors,

$$n_0 = N_c \exp[-(E_G - E_{F0})/kT] \tag{3.35}$$

and

$$p_0 = N_v \exp[-(E_{F0} - E_v)/kT] \tag{3.36}$$

The imperfection occupancy under illumination is given by

$$n_i/N_i = (\beta_{in}n + P_{iv})/(\beta_{in}n + \beta_{ip}p + P_{ic} + P_{iv}) \tag{3.37}$$

Figure 3.3. Energy band diagram for a multilevel imperfection model, including both capture and thermal excitation processes for each level.

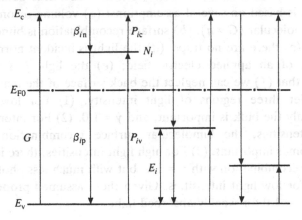

Finally the continuity equation for holes gives

$$G = \sum_i n_i \beta_{ip} p - \sum_j (N_j - n_j) P_{jv} + \beta_{cv} np \qquad (3.38)$$

## 3.4    Model for sublinear behavior without traps

One-center recombination models, i.e., models with only one value of electron and hole capture recombination coefficient, $\beta_n$ and $\beta_p$, result in variations $\Delta n \propto G^\gamma$ with $\gamma = 0.50$ or $1.0$. Experimentally a wide variety of values of $\gamma$ are observed with values between 0.50 and 1.0, as well as values larger than 1.0. The question arises as to what are the simplest variations on a one-center model that are capable of producing such values of $\gamma$. Three answers may be given to this question: (a) use a one-center model involving effects of both surface and volume; (b) use a one-center model with the addition of the effects of traps; and (c) use at least a two-center model, where two types of recombination centers are present with different values of $\beta_n$ and $\beta_p$. We discuss the first of these possibilities in this section, the second in the last section of this chapter, and the third in Chapter 5.

Since it is possible to imagine situations in which $\gamma = 0.50$ and $\gamma = 1.0$ in the same material over different ranges of excitation intensity, it might be possible to obtain a situation where $\gamma$ takes on an intermediate value, $0.50 < \gamma < 1.0$, over a range of light intensities in the transition region from a $\gamma = 0.50$ regime to a $\gamma = 1.0$ regime. A model has been developed for such a possibility in which the $\gamma = 0.50$ regime is characteristic of bimolecular recombination at the surface, and the $\gamma = 1.0$ regime is characteristic of monomolecular recombination in the bulk.[1] Since this model involves several interesting aspects of a more general calculation, we describe it here in a little detail.

The model is based on several assumptions: (a) volume recombination is monomolecular ($G \propto n$); (b) surface recombination is bimolecular ($G \propto n^2$); (c) there are no traps; (d) the light is incident normal to the direction of an applied electric field; (e) the light is strongly absorbed, so that (f) we can neglect the back surface of the material.

We consider three regions of light intensity. (1) For low light intensities, only the bulk is important, and $\gamma = 1.0$. (2) For intermediate light intensities, the bimolecular surface recombination with $\gamma = 0.50$ becomes important. (3) For high light intensities, there is very high surface recombination with $\gamma = 1.0$, but with much less photosensitivity than for low light intensities. Given these assumed properties, the value of $\gamma = 1.0$ for weakly absorbed light.

In steady state

$$G = n/\tau_n = p/\tau_p \qquad (3.39)$$

In the bulk the electron and hole lifetimes, $\tau_n$ and $\tau_p$ are constant, independent of $n$ and $p$. Both electrons and holes diffuse away from the surface, and the particle current perpendicular to the surface is described by

$$\mathbf{J}_n = -n\mu_n \mathscr{E}_D - D_n \, dn/dx \qquad (3.40)$$

$$\mathbf{J}_p = p\mu_p \mathscr{E}_D - D_p \, dp/dx \qquad (3.41)$$

Here $\mathscr{E}_D$ is an electric field set up normal to the surface because of the different transport properties of electrons and holes. This field accelerates the carriers with smaller range and decelerates the carriers with larger range, so that in steady state $\mathbf{J}_n = \mathbf{J}_p$; $\mathscr{E}_D$ is called the 'Dember field.' The Dember field is different from zero under these conditions of strongly absorbed light as long as $r_n = \mu_n \tau_n \neq r_p = \mu_p \tau_p$.

If we subtract Eq. (3.41) from Eq. (3.40), we obtain

$$-n\mathscr{E}_D[\mu_n + (\tau_p/\tau_n)\mu_p] = (kT/q)(dn/dx)[\mu_n - (\tau_p/\tau_n)\mu_p] \qquad (3.42)$$

since $p/n = \tau_p/\tau_n$, which enables us to solve for the Dember field,

$$\mathscr{E}_D = (kT/q)(dn/dx)[(r_p - r_n)/(r_p + r_n)](1/n) \qquad (3.43)$$

Then the electron current density becomes

$$\mathbf{J}_n = -D_n(dn/dx)[2r_p/(r_p + r_n)] = \mathbf{J}_p = -D_p(dp/dx)[2r_n/(r_p + r_n)] \qquad (3.44)$$

The steady-state continuity equations become

$$G - n/\tau_n = \nabla \cdot \mathbf{J}_n = -D_n[2r_p/(r_p + r_n)](d^2n/dx^2) \qquad (3.45)$$

$$G - p/\tau_p = \nabla \cdot \mathbf{J}_p = -D_p[2r_n/(r_p + r_n)](d^2p/dx^2) \qquad (3.46)$$

with

$$G = \alpha F \exp(-\alpha x) = (F/d) \exp(-x/d) \qquad (3.47)$$

if $F$ is the photon flux per unit area per second incident on the material, and $d = 1/\alpha$ is the penetration depth of the light.

Define an ambipolar diffusion length $L$, such that

$$L^2 = (kT/q)[2r_n r_p/(r_n + r_p)] \qquad (3.48)$$

If $r_n = \mu_n \tau_n = qL_n^2/kT$ and $r_p = \mu_p \tau_p = qL_p^2/kT$, then $2/L^2 = 1/L_n^2 + 1/L_p^2$.

Eqs. (3.45) and (3.46) now become

$$L^2(d^2n/dx^2) - n + (F\tau_n/d)\exp(-x/d) = 0 \qquad (3.49)$$

$$L^2(d^2p/dx^2) - p + (F\tau_p/d)\exp(-x/d) = 0 \qquad (3.50)$$

with a solution of the form $n = A\exp(-x/L) - K\exp(-x/d)$. Here $K$ is given by

$$K = F(\tau_n/d)/[(L/d)^2 - 1] \to 0 \quad \text{when } L/d \gg 1 \qquad (3.51)$$

$A$ can be evaluated by setting the rate of surface recombination $C_s n^2(0)$ equal to the rate of pair diffusion to the surface $(L^2/\tau_n)(dn/dx)_{x=0}$, which gives

$$A = (L/C_s\tau_n)((F\tau_n^2 C_s)/\{Ld[(L/d)^2 - 1]\}$$
$$- \tfrac{1}{2} + \{\tfrac{1}{4} + (F\tau_n^2 C_s)/[Ld(L/d + 1)]\}^{1/2}) \qquad (3.52)$$

which also simplifies when $L/d \gg 1$ for strongly absorbed light.

Now the photoconductance per square of illuminated surface is

$$\Gamma = q\mu_n \int_0^\infty n(x)\,dx + q\mu_p \int_0^\infty p(x)\,dx \qquad (3.53)$$

If we neglect the contribution due to holes and the $K$ term in $n(x)$ according to Eq. (3.51), then since $\int_0^\infty A\exp(-x/L)\,dx = AL$, $\Gamma = q\mu_n AL$. If we write this out, using the simplification of Eq. (3.52) associated with $L/d \gg 1$, we have finally

$$\Gamma = (q\mu_n L^2/C_s\tau_n)\{(F\tau_n^2 C_s)/[L^2(L/d)] - \tfrac{1}{2} + [\tfrac{1}{4} + (F\tau_n^2 C_s)/L^2]^{1/2}\} \quad (3.54)$$

We can now explore the significance of Eq. (3.54) in the various ranges of light intensity. For low light intensities, $F \le F_{lo}$, defined by $F_{lo}\tau_n^2 C_s/L^2 = \tfrac{1}{4}$, so that $F_{lo} = L^2/(4\tau_n^2 C_s)$, we neglect the first term since $L/d \gg 1$, and

$$\Gamma_{lo} = (q\mu_n L^2/C_s\tau_n)\{-\tfrac{1}{2} + \tfrac{1}{2}[1 + (4F\tau_n^2 C_s)/L^2]^{1/2}\} \qquad (3.55)$$

If we approximate the square root $(1 + x)^{1/2} = 1 + x/2$ for small $x$ (here small $F$), then

$$\Gamma_{lo} = (q\mu_n L^2/C_s\tau_n)(F\tau_n^2 C_s/L^2) = q\mu_n F\tau_n \qquad (3.55)$$

and $\Gamma_{lo} \propto F$, $\gamma_{lo} = 1.0$.

For intermediate light intensities, $F_{lo} < F < F_{hi}$. Here $F_{hi}$ is defined by the equality between the first term on the right of Eq. (3.54) and the final square root term, which dominates in the intermediate intensity range.

$$(F_{hi}\tau_n^2 C_s)[L^2(L/d)] = (F_{hi}\tau_n^2 C_s/L^2)^{1/2} \qquad (3.57)$$

so that

$$F_{hi} = (L/d)^2 L^2/(\tau_n{}^2 C_s) = 4(L/d)^2 F_{lo} \qquad (3.58)$$

If in this intermediate range we neglect the first term on the right of Eq. (3.54) and the $-\frac{1}{2}$, then

$$\Gamma_{int} = (q\mu_n L^2/C_s\tau_n)(F\tau_n{}^2 C_s/L^2)^{1/2} = q\mu_n(FL^2/C_s)^{1/2} \qquad (3.59)$$

and $\Gamma_{int} \propto F^{1/2}$, $\gamma_{int} = 0.50$.

Finally for high light intensities, $F > F_{hi}$, the first term on the right of Eq. (3.54) dominates,

$$\Gamma_{hi} = (q\mu_n L^2/C_s\tau_n)\{(F\tau_n{}^2 C_s)[L^2(L/d)]\} = q\mu_n F\tau_n/(L/d) \qquad (3.60)$$

and $\Gamma_{hi} \propto F$, $\gamma_{hi} = 1.0$. The photosensitivity in the linear high-intensity range has been reduced by a factor of $L/d$ compared to the photosensitivity in the linear low-intensity range.

Eq. (3.58) shows that there is a range of intensities between $F_{lo}$ and $F_{hi}$, the width of which depends directly on $(L/d)^2$. For a value of $L/d = 30$, for example, there is a range of almost three orders of magnitude over which $\gamma = 0.65$. Figure 3.4 shows a $\Gamma$ vs $F$ plot for typical parameters.

Figure 3.4. Plot of Eq. (3.54) for conductance $\Gamma$ vs photon flux $F$, for $\mu_n = 1 \text{ m}^2 \text{ V}^{-1}\text{s}^{-1}$, $L = 3 \times 10^{-6} \text{ m}$, $d = 10^{-7} \text{ m}$, $C_s = 10^{-15} \text{ m}^3 \text{s}^{-1}$, and $\tau_n = 10^{-9} \text{ s}$. $F_{lo} = 2 \times 10^{21} \text{ m}^{-2} \text{s}^{-1}$, and $F_{hi} = 8 \times 10^{24} \text{ m}^{-2} \text{s}^{-1}$.

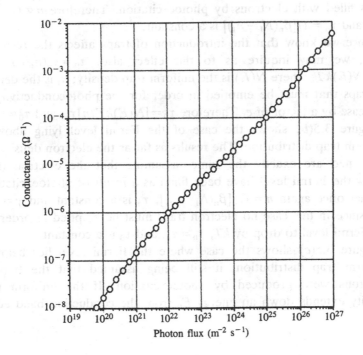

Although consideration of this model has been useful to introduce the Dember potential and to see how calculations of this type can be done, it is clearly not an adequate answer to the general question of why experimental values of $\gamma$ between 0.50 and 1.0 are routinely seen over wide orders of magnitude, for light with small absorption constant, and under situations in which a transition from $\gamma = 0.50$ to $\gamma = 1.0$ is not involved.

## 3.5    One-center model with traps

We have already seen that the introduction of traps into a semiconductor can decrease the photosensitivity. It is reasonable to inquire, therefore, what effects the introduction of traps can have on the expected values of $\gamma$.[2] A number of possible situations are illustrated in Figure 3.5. For the sake of this discussion, we assume that $\tau_p \ll \tau_n$, so that we may restrict our attention to photoconductivity associated with electrons. We also assume that electron trapping dominates, and that we are in the insulator case with $n = \Delta n \gg n_0$.

Figure 3.5(*a*) shows the first case of the Fermi level lying in the midst of a uniform trap distribution, i.e., a distribution of traps in which the density is independent of trap depth. The photoexcited electron density is given by $n = G/[\beta_n(N_I - n_I)]$, with the density of holes in recombination centers $N_I - n_I = N_t$, where $N_t$ is the density of traps filled with electrons by photoexcitation. Therefore $n \propto G$, $\gamma = 1.0$, and $\tau_n = 1/[\beta_n(N_I - n_I)]$ is a constant.

Since we know that the introduction of traps affects the response time, we may inquire as to this effect also. $\tau_0 = \tau_n(n_t/n)$ with $n_t = N(E)kT$, where $N(E)$ is the uniform trap density; $n_t$ is the density of traps that must be emptied in order for the photoconductivity to decrease by a factor of $e$. Therefore $\tau_0 = [N(E)kT/n]\tau_n$, and $\tau_0 \propto G^{-1}$.

Figure 3.5(*b*) shows the case of the Fermi level lying above a uniform trap distribution. The results as far as the electron density are concerned are exactly the same, assuming that the electron traps below the Fermi level have been filled as a result of photoexcitation, so that once again $n = G/[\beta_n(N_I - n_I)]$, $\tau_n$ is a constant and $\gamma = 1.0$. But since in this case no electron traps must be emptied in order for the Fermi level to drop by $kT$, $\tau_0 = \tau_n$, and $\tau_0$ is a constant.

Figure 3.5(*c*) shows the case where the Fermi level lies below a uniform trap distribution, it still being assumed that the trapped electrons were produced by photoexcitation. If the uniform trap density extends down an energy $E_0$ from the conduction band edge, then the total density of trapped electrons is

$$N_t = \int N(E) \exp(-E/kT) \, dE \qquad (3.61)$$

where $N(E)$ is the uniform trap density, the lower limit for the integration is $(E_c - E_0 - E_{Fn})$ and the upper limit is $(E_c - E_{Fn})$. The result is

$$N_t \approx kTN(E) \exp[-(E_c - E_{Fn})/kT] \exp(E_0/kT) \qquad (3.62)$$

under the assumption that $\exp(E_0/kT) \gg 1$. We may therefore write $N_t = \alpha(T)n$, with $\alpha(T) = [kTN(E) \exp(E_0/kT)/N_c]$. If we assume that

Figure 3.5. Energy band diagrams for considering the effect of traps on photoconductivity magnitude and speed of response. (a) Fermi level in the midst of a uniform trap distribution, (b) Fermi level above a uniform trap distribution, (c) Fermi level below a uniform trap distribution, (d) Fermi level between two uniform trap distributions, and (e) Fermi level in the midst of an exponential trap distribution.

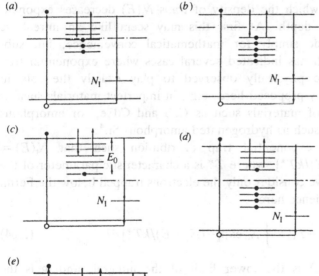

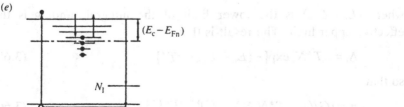

$N_t \gg n$, then $(N_I - n_I) \approx N_t$, and

$$n = \{G/[\beta_n\alpha(T)]\}^{1/2} \quad \text{so that} \quad n \propto G^{1/2} \tag{3.63a}$$

$$\tau = \{1/[\beta_n\alpha(T)G]\}^{1/2} \quad \text{so that} \quad \tau \propto G^{-1/2} \tag{3.63b}$$

$$\tau_0 = \tau_n(N_t/n) = \alpha(T)\tau_n \quad \text{so that} \quad \tau_0 \propto G^{-1/2} \tag{3.63c}$$

If the Fermi level lies in a void between two uniform trap distributions, such as is shown in Figure 3.5(d), then the behavior will depend on whether there are more photoexcited electrons trapped below or above the Fermi level. If there are more electrons trapped below the Fermi level, then the variation of $n$ with $G$ is like Figure 3.5(b) but the variation of $\tau_0$ is like Figure 3.5(c): $n \propto G$ and $\tau_0 = \alpha(T)\tau_n$. If there are more electrons trapped above the Fermi level, then the behavior is like Figure 3.5(c): $n \propto G^{1/2}$ and $\tau_0 \propto G^{-1/2}$.

In all of the above cases involving a uniform trap distribution, the only values of $\gamma$ that are found are still either 0.50 or 1.0. It appears, therefore, that the addition of a uniform distribution of traps to the model is not able to achieve values of $\gamma$ between 0.50 and 1.0. A situation in which the density of traps varies with depth appears to be indicated. Figure 3.5(e) shows a particular case of an exponential trap distribution, in which the density of traps $N_t(E)$ decreases exponentially with trap depth. At first this may seem like an unrealistic assumption made simply for mathematical convenience, but subsequent research has indicated several cases where exponential trap distributions are physically observed to play exactly the role in photoconductivity proposed here, e.g., in imperfect materials such as sintered layers of materials such as CdS and CdSe,[3] or amorphous semiconductors such as hydrogenated amorphous Si.[4]

Consider an exponential trap distribution such that $N_t(E) = N_0 \exp[-(E_c - E)/kT^*]$, where $T^*$ is a characteristic parameter of the distribution. If we consider only the electrons trapped below the Fermi level for convenience here,

$$N_t \approx (N_I - n_I) \approx \int N_0 \exp[-(E_c - E)/kT^*] \, dE \tag{3.64}$$

where $(E_c - E_{Fn})$ is the lower limit of the integral, and $\infty$ is the effective upper limit. The result is that

$$N_t = kT^*N_0 \exp[-(E_c - E_{Fn})/kT^*] \tag{3.65}$$

so that

$$n = [G/(\beta_n kT^*N_0N_c)^{-T/T^*}]^{[T^*/(T+T^*)]} \tag{3.66}$$

since $\exp[-(E_c - E_{Fn})/kT^*] = (n/N_c)^{T/T^*}$. The electron lifetime is given by

$$\tau_n \propto G^{[-T/(T+T^*)]} \tag{3.67}$$

since $\tau_n \propto n^{-(T/T^*)}$. The response time $\tau_0 = \tau_n(n_t/n)$, where $n_t = kT N_0 \exp[-(E_c - E_{Fn})/kT^*]$, so that

$$\tau_0 \propto G^{[-T^*/(T+T^*)]} \tag{3.68}$$

The specific behavior and value of $\gamma$ observed therefore depend on the relative magnitudes of $T$ and $T^*$. If $T^* \gg T$, then $n \propto G$ since we effectively have a uniform distribution. If $T^* = T$, then we have the special case of $\gamma = 0.50$. If $T^* < T$, then the above calculation is really not valid since the density of trapped electrons below the Fermi level is less than the density of electrons trapped above the Fermi level, and $\gamma = 0.50$. But in the remaining cases in which $T^* \geq T$, $n \propto G^\gamma$ with $0.50 < \gamma < 1.0$.

Clearly it is not necessary for an ideal exponential trap distribution to exist over the whole bandgap region for this kind of behavior to be observed experimentally. It is required only that the variation of traps with depth be exponential over the range of variation of the steady state Fermi level caused by the photoexcitation intensity range.

# 4

# The Shockley–Read one-center model

The most general one-center model, characterized by having a single type of recombination center with capture coefficients $\beta_n$ and $\beta_p$, and with thermal excitation processes between the level and both the conduction and the valence bands, has been analyzed in some detail by Shockley and Read.[1] This model has been so widely used in a variety of applications to semiconductors, that special attention to it is merited here. A one-level model is more general than might first appear; whenever the recombination kinetics are controlled over a specific range of temperature and light intensity by imperfections of one type, a one-center model is adequate for that range of variables.

## 4.1    General properties of the model

The general properties of the model are illustrated in Figure 4.1. In describing the model, we will stay as close as possible to the notation used in the original published derivation. An imperfection level lies $E_I$ above the valence band edge, which in this modeling we will take as the zero point energy (i.e., $E_v = 0$). Under a photogeneration rate $G$ across the bandgap, free electron densities $n = n_0 + \Delta n$, and $p = p_0 + \Delta p$ exist, where the "0" subscript indicates the dark values. Electrons are captured by the imperfections with a rate $R_n$ and holes with a rate $R_p$. Thermal excitation of electrons out of the imperfection level to the conduction band, and of holes out of the imperfection level to the valence band are included in the model.

In steady state $G = R_n = R_p$, and $\tau_n = \Delta n/G$, $\tau_p = \Delta p/G$. Explicit values for $R_n$ and $R_p$ can be given:

$$dn_I/dt = R_n = n(N_I - n_I)\beta_n - n_I N_c \beta_n \exp[-(E_G - E_I)/kT] \tag{4.1}$$

$$d(N_I - n_I)/dt = R_p = n_I p \beta_p - (N_I - n_I)N_v \beta_p \exp(-E_I/kT) \tag{4.2}$$

It is convenient to separate the solution into four cases, each more general than the preceding:

(A) $\quad \Delta n, \Delta p \ll n_0, p_0 \qquad \tau_n = \tau_p \qquad \Delta n = \Delta p$ (4.3a)

(B) $\quad \Delta n, \Delta p \gtrsim n_0, p_0 \qquad \tau_n = \tau_p \qquad \Delta n = \Delta p$ (4.3b)

(C) $\quad \Delta n, \Delta p \ll n_0, p_0 \qquad \tau_n \neq \tau_p \qquad \Delta n \neq \Delta p$ (4.3c)

(D) $\quad \Delta n, \Delta p \gtrsim n_0, p_0 \qquad \tau_n \neq \tau_p \qquad \Delta n \neq \Delta p$ (4.3d)

To follow the traditional notation, set $n_I = f_I N_I$, so that $(N_I - n_I) = (1 - f_I)N_I$, and set $n' = N_c \exp[-(E_G - E_I)/kT]$, and $p' = N_v \exp(-E_I/kT)$. We see that $n'$ is the value of $n$, and that $p'$ is the value of $p$ when the Fermi level lies at $E_I$. (For correspondence with our former notation in Eqs. (3.31) and (3.32), note that $n'\beta_n = P_c$ and $p'\beta_p = P_v$.) Then Eqs (4.1) and (4.2) can be rewritten:

$$R_n = n N_I(1 - f_I)\beta_n - N_I f_I n'\beta_n$$ (4.4)

$$R_p = p N_I f_I \beta_p - N_I(1 - f_I)p'\beta_p$$ (4.5)

If $R_n = R_p$,

$$f_I = (\beta_n n + \beta_p p')/[\beta_n(n + n') + \beta_p(p + p')]$$ (4.6)

which is equivalent to Eq. (3.37). Substituting back gives

$$G = (pn - p'n')/[(n + n')/C_p + (p + p')/C_n]$$ (4.7)

Figure 4.1. Energy band diagram for the Shockley–Read recombination model. A recombination center with density $N_I$, characterized by capture coefficients $\beta_n$ and $\beta_p$, lies $E_I$ above the valence band.

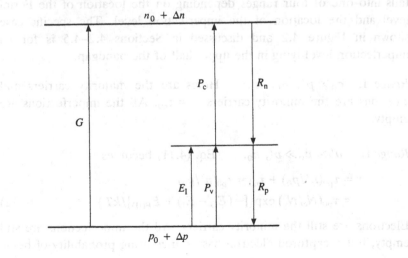

where we have set $C_n = N_I \beta_n$ and $C_p = N_I \beta_p$. Often $1/C_n$ is set equal to $\tau_{n0}$, which represents the minimum lifetime for an electron when all the imperfections are empty, and $1/C_p$ is set equal to $\tau_{p0}$, which represents the minimum lifetime for a hole when all the imperfections are electron occupied. We may also note that $n'p' = n_0 p_0 = n_i^2$, where $n_i$ is the intrinsic free carrier density.

Expressing explicitly the effect of photoexcitation, Eq. (4.7) becomes

$$G = [(p_0 + \Delta p)(n_0 + \Delta n) - p'n']/[(n_0 + \Delta n + n')/C_p$$
$$+ (p_0 + \Delta p + p')/C_n] \quad (4.8)$$

or

$$G = (\Delta p n_0 + \Delta n p_0 + \Delta n \Delta p)/[(n_0 + \Delta n + n')/C_p + (p_0 + \Delta p + p')/C_n]$$
$$(4.9)$$

This is the fundamental equation from which specific results for the lifetimes for each of the four cases enumerated above can be derived.

## 4.2    Case A: small signals, equal lifetimes

Under the conditions of Eq. (4.3a)

$$G = \Delta n(n_0 + p_0)/[(n_0 + n')/C_p + (p_0 + p')/C_n] \quad (4.10)$$

and we can immediately write the lifetime as

$$\tau = \Delta n/G = \tau_{p0}[(n_0 + n')/(n_0 + p_0)] + \tau_{n0}[(p_0 + p')/(n_0 + p_0)] \quad (4.11)$$

A typical result for $\tau$ vs $E_F$ is pictured in Figure 4.2(a) with a corresponding energy band diagram in Figure 4.2(b). The solution falls into one of four ranges depending on the location of the Fermi level and the location of the imperfection level. The specific case shown in Figure 4.2 and discussed in Sections 4.2–4.5 is for an imperfection level lying in the upper half of the bandgap.

*Range* I: $p_0 \gg p'$, $n_0$, $n'$.    Holes are the majority carriers and electrons are the minority carriers. $\tau = \tau_{n0}$. All the imperfections are empty.

*Range* II: $n' \gg p_0 \gg p'$, $n_0$.    Eq. (4.11) becomes

$$\tau = \tau_{p0}(n'/p_0) + \tau_{n0} \approx \tau_{p0}(n'/p_0)$$
$$= \tau_{p0}(N_c/N_v) \exp\{[-(E_G - E_I) + E_{F(II)}]/kT\} \quad (4.12)$$

Electrons are still the minority carriers. All the imperfections are still empty, but a captured electron has an increasing probability of being

Figure 4.2. Results of Shockley–Read model for small signal, equal
lifetime case, described by Eq. (4.11). (a) Typical variation of
lifetime with Fermi level position. Calculated in this case for
$\tau_{p0} = 10^{-9}$ s, $\tau_{n0} = 10^{-8}$ s, $E_I = 0.8$ eV, $E_G = 1.0$ eV, $N_c = 2 \times 10^{18}$ cm$^{-3}$, and $N_v = 8 \times 10^{18}$ cm$^{-3}$. (b) Separation of the bandgap into
four regions corresponding to data of (a): I ($0 < E_F < 0.2$), II
($0.2 < E_F < 0.5$), III ($0.5 < E_F < 0.8$), IV ($0.8 < E_F < 1.0$).

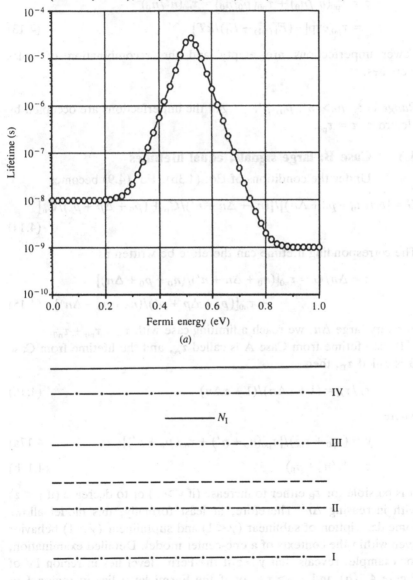

(a)

(b)

thermally re-excited to the conduction band rather than recombining with a hole.

*Range* III: $n' \gg n_0 \gg p_0 \gg p'$. Now electrons are the majority carriers, and holes are the minority carriers. Eq. (4.11) becomes

$$\tau = \tau_{p0}(n'/n_0) + \tau_{n0}(p_0/n_0) \approx \tau_{p0}(n'/n_0)$$
$$= \tau_{p0} \exp[-(E_{F(III)} - E_I)/kT] \tag{4.13}$$

Fewer imperfections are empty and the recombination of holes increases.

*Range* IV: $n_0 \gg n'$, $p_0$, $p'$. All the imperfections are occupied by electrons. $\tau = \tau_{p0}$.

## 4.3 Case B: large signals, equal lifetimes

Under the conditions of Eq. (4.3b), Eq. (4.9) becomes

$$G = [\Delta n(n_0 + p_0 + \Delta n)]/[(n_0 + \Delta n + n')/C_p + (p_0 + \Delta p + p')/C_n] \tag{4.14}$$

The corresponding lifetime can therefore be written as

$$\tau = \Delta n/G = \tau_{p0}[(n_0 + \Delta n + n')/(n_0 + p_0 + \Delta n)]$$
$$+ \tau_{n0}[(p_0 + \Delta p + p')/(n_0 + p_0 + \Delta n)] \tag{4.15}$$

For very large $\Delta n$, we reach a limiting case with $\tau_\infty = \tau_{p0} + \tau_{n0}$.

If the lifetime from Case A is called $\tau_A$, and the lifetime from Case B is called $\tau_B$, then

$$\tau_B/\tau_A = (1 + y\Delta n)/(1 + z\Delta n) \tag{4.16}$$

where

$$y = (\tau_{p0} + \tau_{n0})/[\tau_{p0}(n_0 + n') + \tau_{n0}(p_0 + p')] \tag{4.17a}$$
$$z = 1/(n_0 + p_0) \tag{4.17b}$$

It is possible for $\tau_B$ either to increase (if $y > z$) or to decrease (if $y < z$) with increasing $\Delta n$. Therefore, at least formally, this model allows some description of sublinear ($\gamma < 1$) and supralinear ($\gamma > 1$) behavior even within the contexts of a one-center model. Detailed examination, for example, reveals that $y > z$ if the Fermi level lies in region IV of Figure 4.2(b) and $\tau_{n0} > \tau_{p0}$, or if the Fermi level lies in region I of Figure 4.2(b) and $\tau_{n0} < \tau_{p0}$. Similarly, if the Fermi level lies in either regions II or III of Figure 4.2(b), and $\tau_{n0} < \tau_{p0}$, $y < z$.

## 4.4 Case C: small signals, unequal lifetimes

Under the conditions of Eq. (4.3c), Eq. (4.9) becomes

$$G = (\Delta p n_0 + \Delta n p_0)/[(n_0 + \Delta n + n')/C_p + (p_0 + \Delta p + p')/C_n] \quad (4.18)$$

Since for this case $\Delta n \neq \Delta p$, it is necessary to calculate the photoexcited change in the occupation function $\Delta f_I$. To do this, rewrite Eq. (4.4):

$$R_n = (n_0 + \Delta n)N_I(1 - f_{I0} - \Delta f_I)\beta_n - N_I(f_{I0} + \Delta f_I)n'\beta_n$$
$$= R_{n0} + C_n[(1 - f_{I0})\Delta n - (n_0 + n')\Delta f_I] \quad (4.19)$$

Similarly Eq. (4.5) can be rewritten:

$$R_p = R_{p0} + C_p[f_{I0}\Delta p + (p_0 + p')\Delta f_I] \quad (4.20)$$

Setting $R_n = R_p$ gives

$$\Delta f_I = [\tau_{p0}(1 - f_{I0})\Delta n - \tau_{n0}f_{I0}\Delta p]/[\tau_{n0}(p_0 + p') + \tau_{p0}(n_0 + n')] \quad (4.21)$$

It is necessary to eliminate $\Delta n$ or $\Delta p$ to calculate $\tau_p$ or $\tau_n$ using $\Delta p - \Delta n = N_I \Delta f_I$. For example, in order to calculate $\tau_p$, (i) substitute $\Delta n = \Delta p - N_I \Delta f_I$ into Eq. (4.21) and solve for $\Delta f_I(\Delta p)$, (ii) substitute $\Delta n = \Delta p - N_I \Delta f_I$ into Eq. (4.18) and solve for $G(\Delta p, \Delta f_I)$, (iii) substitute for $\Delta f_I$, and (iv) calculate $\tau_p = \Delta p/G$. The result is

$$\tau_p = \{\tau_{n0}(p_0 + p') + \tau_{p0}[n_0 + n' + N_I(1 - f_{I0})]\}/[n_0 + p_0 + N_I f_{I0}(1 - f_{I0})] \quad (4.22)$$

Similarly one can obtain

$$\tau_n = \{\tau_{p0}(n_0 + n') + \tau_{n0}[p_0 + p' + N_I f_{I0}]\}/[n_0 + p_0 + N_I f_{I0}(1 - f_{I0})] \quad (4.23)$$

## 4.5 Case D: large signals, unequal lifetimes

This is the most general case and most difficult for which to produce an exact analytical solution. A close approximation to the general solution has been reported:[2]

$$\tau_n \approx \frac{\begin{aligned}\tau_{p0}(n_0 + n' + \Delta n) + \tau_{n0}(p_0 + p' + \Delta n) \\ + \tau_{n0}N_I[p'(n_0 + n' + \Delta n) + 2p_0\Delta n]/[(n_0 + n' + \Delta n)(p_0 + p')]\end{aligned}}{(n_0 + p_0 + \Delta n) + N_I[p_0(n_0 + \Delta n)]/[(n_0 + n' + \Delta n)(p_0 + p')]} \quad (4.24)$$

$$\tau_p \approx \frac{\begin{aligned}\tau_{p0}(n_0 + n' + \Delta p) + \tau_{n0}(p_0 + p' + \Delta p) \\ + \tau_{p0}N_I[(p_0 + p' + \Delta p) + 2p'\Delta p]/[(p_0 + p' + \Delta p)(p_0 + p')]\end{aligned}}{(n_0 + p_0 + \Delta p) + N_I[p'(p_0 + \Delta p)]/[(p_0 + p' + \Delta p)(p_0 + p')]} \quad (4.25)$$

As a specific example of the application of these results, consider a case where the Fermi level lies above the imperfection level, which lies above the middle of the gap, and $\Delta n \gg n_0$. Then the relationship between the various densities is given by $\Delta n > n_0 > n' > p' > p_0$, and Eq. (4.24) reduces simply to

$$\tau_n = (\tau_{p0} + \tau_{n0})\Delta n + \tau_{n0}N_I/[\Delta n + N_I(p_0/p')] \qquad (4.26)$$

Suppose for the sake of this example that $\tau_{p0} \ll \tau_{n0}$. Then Eq. (4.26) becomes

$$\tau_n = \tau_{n0}\{(\Delta n + N_I)/[\Delta n + N_I(p_0/p')]\} \qquad (4.27)$$

The variation of $\Delta n$ with photoexcitation rate $G$ is given for several temperatures in Figure 4.3.

Three ranges are indicated. If $\Delta n < N_I(p_0/p') < N_I$,

$$\tau_n = \tau_{n0}(p'/p_0) = \tau_{n0}\exp[(E_F - E_I)/kT] \qquad (4.28)$$

Figure 4.3. Illustration of the behavior indicated by Eq. (4.27) for a particular simplification of the Shockley–Read recombination expression. Specifically $\tau_{n0} = 10^{-8}$ s, $N_I = 10^{17}$ cm$^{-3}$, $E_F = 1.0$ eV, $E_I = 0.8$ eV, and $N_v = 8 \times 10^{18}$ cm$^{-3}$. From top to bottom, the curves are for 200 K, 300 K, 400 K and 600 K.

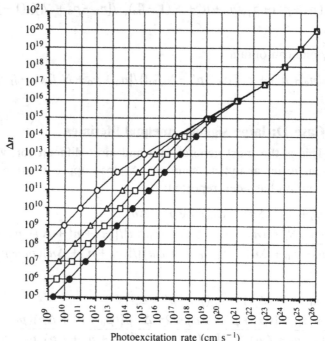

Photoexcitation rate (cm s$^{-1}$)

Over the range of $G$ corresponding to this condition, the lifetime is a constant at a fixed temperature and decreases with increasing temperature. If $N_I(p_0/p') < \Delta n < N_I$,

$$\tau_n = \tau_{n0}(1 + N_I/\Delta n) = \tau_{n0}(N_I/\Delta n) = (\tau_{n0}N_I/G)^{1/2} \quad (4.29)$$

Over the range of $G$ corresponding to this condition, the lifetime decreases as $G^{-1/2}$ and is independent of temperature. Finally if $\Delta n > N_I$, $\tau_n = \tau_{n0}$. For these high values of $G$, the lifetime is once again constant and independent of temperature.

## 4.6 Comparison with computer model calculations

In order to compare the effectiveness of the general computer model of Section 3.3 with the analytical results described above for the Shockley–Read recombination model, we compare results of the two calculations.[3] Figure 4.4 shows the results for the variation of electron and hole density as a function of photoexcitation rate for particular values of the parameters for $10^{15}$ cm$^{-3}$ acceptors with ionization energy of 0.6 eV in a semiconductor with bandgap of 1.0 eV at 100 K, for different values of the dark Fermi energy. Apparent small discrepancies between the model calculation and the analytical calculation can be attributed to the small errors in the approximate analytical calculations underlying Eqs. (4.24) and (4.25).[2] Within the limits of the requirement that $\Delta n \geq n_0$, the model reproduces all of the details of the analytical Shockley–Read model.

An interesting feature of the results of Figure 4.4 is that both the electron and hole lifetimes exhibit a maximum for a particular value of the Fermi energy for low photoexcitation intensities. For the parameters used, the analytical expression for the electron lifetime is

$$\tau_n = \tau_{n0}N_I/[n_0 + N_I(p_0/p')] = n_0p'\tau_{n0}N_I/(n_0^2p' + n_i^2N_I) \quad (4.30)$$

if we write $p_0 = n_i^2/n_0$. The maximum value of Eq. (4.30) occurs when $n_0^2p' = n_i^2N_I$, or when $n_0 = (n'N_I)^{1/2} = 3.6 \times 10^6$ cm$^{-3}$. Similarly the analytical expression for the hole lifetime is

$$\tau_p = \tau_{p0}N_I/[p_0 + N_I(p'/p_0)] = p_0\tau_{p0}N_I/(p_0^2 + p'N_I) \quad (4.31)$$

which has a maximum value for $p_0 = (p'N_I)^{1/2} = 71.5$ cm$^{-3}$, which corresponds to a value of $E_F = 0.34$ eV, in agreement with Figure 4.4(b).

## 4.7 Two Shockley–Read recombination centers

If the recombination rates through two different imperfections were independent, then we would expect that the recombination rates would add and the total lifetime would be given by

$$1/\tau = 1/\tau_1 + 1/\tau_2 \tag{4.32}$$

where $\tau_1$ is the Shockley–Read lifetime for the first imperfection alone, and $\tau_2$ is the Shockley–Read lifetime for the second imperfection alone. As usual, if $\tau_1$ and $\tau_2$ were widely different in magnitude, it would be expected that the total lifetime would be approximately

Figure 4.4. Comparison of electron density ($a$) and hole density ($b$) as calculated from Eqs. (4.24) and (4.25) (solid curves) and those calculated from the model of Section 3.3 (data points), for acceptors with $N_A = 10^{15} \, \text{cm}^{-3}$, $E_A = 0.6 \, \text{eV}$, $E_G = 1.0 \, \text{eV}$, $\beta_n = 10^{-16} \, v_n \, \text{cm}^3 \, \text{s}^{-1}$, $\beta_p = 10^{-12}(300/T)^2 v_p \, \text{cm}^3 \, \text{s}^{-1}$ at 100 K, $\beta_{cv} = 10^{-20}(v_n v_p)^{1/2}$, $m_e^* = 0.2m$, and $m_h^* = 0.5m$.[3] Arrows indicate in ($a$) the points at which $n = n_0$ for $E_F = 0.7$, 0.8 and 0.9 eV, and in ($b$) the points at which $p = p_0$ for $E_F = 0.1$, 0.2, and 0.3 eV.

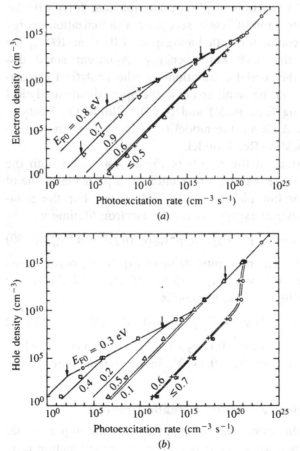

equal to the smaller of the two. This expectation is realized if both imperfections are of the same electrical type, i.e., if both are acceptors or both are donors.

When one imperfection is an acceptor and the other is a donor, however, major departures from Eq. (4.32) may well be found. An analytical solution of the problem is not possible, but results may be obtained by using the general photoconductivity model of Section 3.3. In only a couple of cases is the prediction of Eq. (4.32) confirmed: if the Fermi level lies below the acceptor level, then the electron lifetime

Figure 4.5. Electron lifetime ($a$) and hole lifetime ($b$) as a function of photoexcitation rate for a 0.6 eV donor and a 0.4 eV acceptor only, and a 0.6 eV donor and a 0.4 eV acceptor present simultaneously for $E_F = 0.5$ eV.[3] Calculated for $T = 100$ K, $m_e^* = 0.2m$, $m_h^* = 0.5m$, $N_A = N_D = 10^{15}$ cm$^{-3}$, $\beta_{An} = 10^{-16} v_n$ cm$^3$ s$^{-1}$, $\beta_{Ap} = 10^{-12}(300/T)^2 v_p$ cm$^3$ s$^{-1}$, $\beta_{Dn} = 10^{-12}(300/T)^2 v_n$ cm$^2$ s$^{-1}$, $\beta_{Dp} = 10^{-16} v_p$ cm$^3$ s$^{-1}$, $\beta_{cv} = 10^{-20}(v_n v_p)^{1/2}$ cm$^3$ s$^{-1}$.

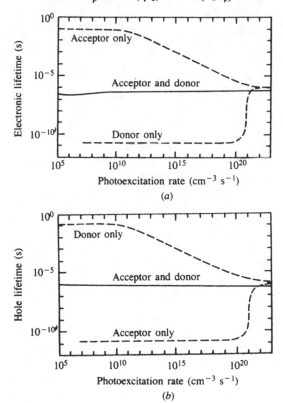

$(a)$

$(b)$

is given correctly by Eq. (4.32), and when the Fermi level lies above the donor level, the hole lifetime is given correctly by Eq. (4.32).

For all other possibilities, however, Eq. (4.32) is no longer reliable. An example of a particularly obvious departure from Eq. (4.32) is shown in Figure 4.5,[3] which has been calculated for a donor lying 0.6 eV above the valence band and an acceptor lying 0.4 eV above the valence band in a semiconductor with bandgap of 1.0 eV, for the case when $E_F = 0.5$ eV. The electron lifetime and hole lifetime are approximately equal to each other, but many orders of magnitude different from any of the four individual Shockley–Read lifetimes over all but very high photoexcitation rates.

# 5

# Two-center recombination effects

The next simplest recombination model to a one-center model is, not surprisingly, a two-center model, i.e., a model in which two recombination centers are present with markedly different capture coefficients for at least one of the carriers. A typical example is given in Figure 5.1, where two recombination centers correspond to levels below the dark Fermi level; for $n$-type effects, the common situation is for $\beta_p^S \approx \beta_p^R$ while $\beta_n^S \ll \beta_n^R$. The superscripts R and S refer to 'recombination' and 'sensitizing' for reasons that will become clear shortly. A more detailed discussion of the physical basis for recombination coefficients is given in Chapter 6. At present it is sufficient to note that the recombination center might be a normal singly negative acceptor, with a Coulomb attractive capture coefficient for holes and a neutral capture coefficient for electrons. The sensitizing center might be a doubly negative acceptor, with a Coulomb attractive capture coefficient for holes (about four times larger than that of the single acceptor) and a much smaller Coulomb repulsive capture coefficient for electrons. The interaction between these two recombination centers gives rise to at least six specific phenomena.

## 5.1    Two-center phenomena

In this section we summarize the qualitative nature of the major phenomena associated with two-center effects and provide some typical examples. The effects considered are (1) imperfection sensitization, (2) supralinear photoconductivity, (3) thermal quenching of photoconductivity, (4) optical quenching of photoconductivity, (5) negative photoconductivity, and (6) photoconductivity saturation. In the following section, we consider the formal details of a two-center model able to describe these effects.

### Imperfection sensitization

This term is used to apply to the situation where the incorporation of additional imperfections leads to a longer lifetime for one of the carriers than was previously present. An example is given in Figure 5.2 for the variation of dark and light conductivity in initially insulating CdS as a function of the density of incorporated halogen (Cl, Br, or I) impurities. As the density of halogen incorporated increases, both the dark and light conductivity increase, but in such a way that a maximum ratio of light to dark conductivity occurs at an intermediate halogen density. The effect can be operated in the reverse direction as well; the inset in Figure 5.2 shows the decrease of dark and light conductivity in a conducting n-type CdS with excess Cd donors when Cu acceptors are added to reduce the conductivity.

The phenomena occurring with halogen incorporation can be pictured in terms of the following chemical effects. Incorporation of a halogen into CdS can proceed by one of two steps: (a) simple substitution of a halogen donor for a S, thus leading to an increase in n-type dark conductivity, and (b) formation of Cd vacancy acceptors by the halogen donors in a process of self-compensation. The results pictured in Figure 5.2 can therefore be described by considering that the lower densities of incorporated halogen donors are involved primarily in the self-compensation formation of Cd vacancies, doubly negatively charged acceptors, which can play the role of sensitizing centers in photoconductivity; thus the light conductivity rises before the dark conductivity. For higher halogen densities, most of the

Figure 5.1. Energy band diagram for a two-center model, with a density $N_R$ of recombination imperfections with capture coefficients $\beta_n{}^R$ and $\beta_p{}^R$, and a density $N_S$ of sensitizing imperfections with capture coefficients $\beta_n{}^S$ and $\beta_p{}^S$.

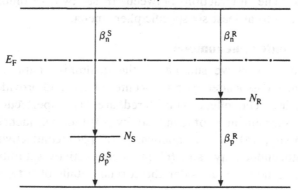

additional halogen simply substitute for S, and the dark conductivity rises. Note that the photosensitivity itself ($\Delta\sigma/G$) increases monotonically with increasing halogen concentration, whereas the ratio of light to dark conductivity ($\Delta\sigma/\sigma_0$) has a maximum at an intermediate halogen concentration.

In the case of incorporation of Cu acceptors into high-conductivity, n-type evaporated CdS layers, we may assume that the material as prepared has a high density of Cd vacancies and hence a high photosensitivity, whereas the incorporation of Cu adds both acceptors to reduce the dark conductivity and recombination centers to compete with the defect sensitizing centers.

### Supralinear photoconductivity

A two-center model is able to describe the interesting situation when it is found that $\Delta n \propto G^\gamma$ with $\gamma > 1.0$, the formal definition of

Figure 5.2. Light conductivity and dark conductivity of initially insulating and undoped CdS as a function of incorporated halide concentration, and (inset)[1] of initially n-type conducting evaporated CdS layers as a function of incorporated Cu concentration.

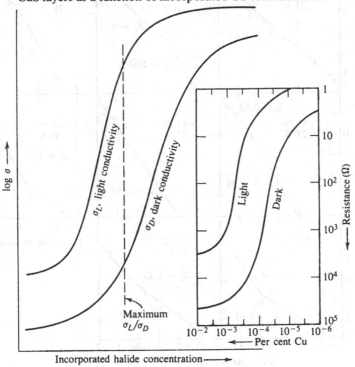

'supralinear' photoconductivity. A typical example is given in Figure 5.3 ($S_2 = \gamma > 1 \cdot 0$ in Figure 5.3), showing the variation of photoconductivity with photoexcitation intensity at several different temperatures for a crystal of CdSe. For the intensities used in these experiments, a range of temperatures exists in which a supralinear portion of the $\Delta \sigma$

Figure 5.3. Photocurrent as a function of photoexcitation intensity for a crystal of CdSe for several different temperatures. The temperatures, applied voltages, and slopes of the different ranges are indicated on the figure.

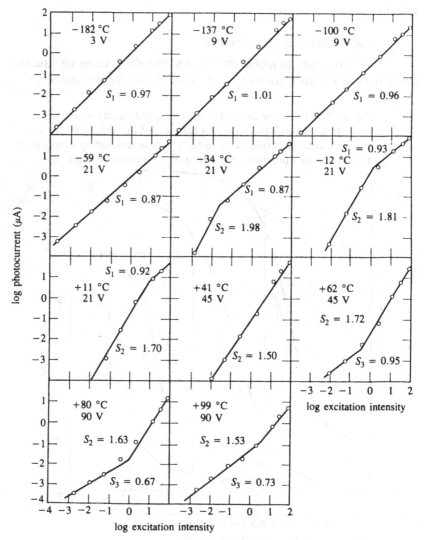

vs $G$ plot is observed. Supralinear photoconductivity corresponds to an increase in lifetime with photoexcitation intensity, i.e., the material is becoming more photosensitive with increasing photoexcitation intensity. Similar data are shown in Figure 5.4(b), where the variation of

Figure 5.4. (a) Photocurrent for a CdSe crystal as a function of temperature for different photoexcitation intensities, and (b) the same data plotted as photocurrent as a function of photoexcitation intensity for different temperatures.

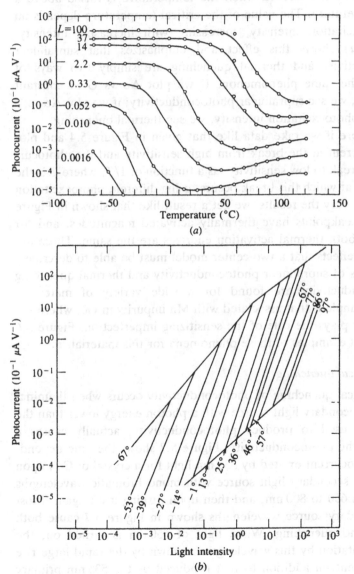

(a)

(b)

photocurrent with photoexcitation intensity at different temperatures is plotted all on the same graph. This shows that the supralinear region is an intermediate transition between two linear or sublinear regions.

### Thermal quenching

When the photoconductivity of a material involving two-center effects is measured as a function of temperature, an abrupt decrease in photosensitivity is observed when the temperature is raised above a critical temperature. The value of the critical temperature depends on the photoexcitation intensity, increasing with increasing intensity. Figure 5.4($a$) shows this effect and emphasizes that supralinear photoconductivity and thermal quenching are simply two ways of looking at the same phenomenon. If we plot $\Delta\sigma$ vs $G$ at constant temperature, we see supralinear photoconductivity; if we plot $\Delta\sigma$ vs $T$ at constant photoexcitation intensity, we see thermal quenching.

Furthermore if we take data like that given in Figure 5.4 and plot the photocurrent at the break from high sensitivity and the photocurrent at the break to low sensitivity, as a function of $1/T$, where $T$ is the temperature at which this break occurs, using different photoexcitation intensities to vary the results, we get a result like that shown in Figure 5.5. Both breakpoints have thermally activated magnitudes, and for this sample both thermal activation energies are the same. These are some of the effects that a two-center model must be able to describe.

The effects of supralinear photoconductivity and thermal quenching of photoconductivity are found for a wide variety of materials. Another example is that associated with Mn impurity in Ge, where the Mn impurity plays the role of the sensitizing imperfection. Figure 5.6 shows typical examples of these phenomena for this material.[2]

### Optical quenching

Optical quenching of photoconductivity occurs when illumination with a secondary light source with a photon energy lower than the light source used to produce photoconductivity, actually causes a decrease in the photoconductivity. Figure 5.7 shows the time dependence of photocurrent excited by 535 nm light for a crystal of CdS upon turning on a secondary light source with monochromatic wavelengths varying from 650 to 800 nm, and then upon turning it off again. Most of the secondary source wavelengths shown in Figure 5.7 cause both excitation and quenching. When light of 650 nm is turned on, the effect of excitation by this wavelength is shown by the rapid large rise in photocurrent, in addition to that produced by the 535 nm primary light source; quenching by this wavelength is shown by the decreased current after turning off the 650 nm light, which recovers slowly to the unquenched value. As the wavelength of the secondary source is increased, the ratio of excitation to quenching decreases, until finally for a secondary wavelength of 800 nm, only quenching is observed.

If the steady state current with both primary and secondary light sources is $I_{p+s}$ and the initial maximum current is $I_{max}$ (if it exceeds the photocurrent $I_p$ due to the primary source alone), we may define 'per cent quenching' as either $100(I_{max} - I_{p+s})/I_{max}$ if the secondary light causes photoexcitation of its own, or $100(I_p - I_{p+s})/I_p$ if it does not.

A plot of per cent quenching as a function of secondary wavelength or photon energy then produces a spectral response curve for optical quenching. Typical curves for a variety of CdS crystals are shown in Figure 5.8. Spectra for a variety of different kinds of CdS share the same features: a narrow quenching band peaked at about 0.9 eV, and a quenching edge at about 1.1 eV with quenching extending to higher energies until overwhelmed by photoexcitation by the same light. The low-energy narrow band disappears at low temperatures, indicating that a thermal excitation step is involved.

Figure 5.5. From the data given in Figure 5.4: the photocurrent at the break from high sensitivity as a function of the temperature at which this break occurs, and the photocurrent at the break to low sensitivity as a function of the temperature at which this break occurs.

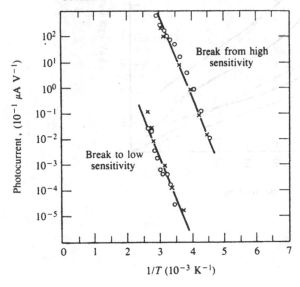

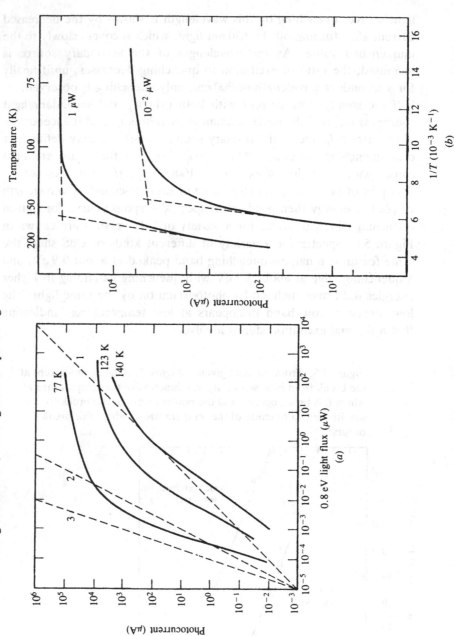

Figure 5.6. Two-center effects in Ge doped with Mn impurity.[2] (a) Photocurrent as a function of light intensity, showing supralinear photoconductivity, and (b) photocurrent as a function of temperature at two different light intensities, showing thermal quenching.

Similar spectra are found for both CdS and ZnS for (*a*) quenching of photoconductivity (as in Figure 5.8), as well as (*b*) photoexcitation of infrared luminescence emission, and (*c*) photoexcitation of hole photoconductivity.[3]

Imperfection sensitization, supralinear photoconductivity, thermal quenching and optical quenching of photoconductivity are the major four phenomena characteristic of two-center recombination. The other two phenomena named above are more exotic, and here we will simply define them and say a few words about them later when our two-center model is defined.

### Negative photoconductivity

It is observed in special materials that illumination can cause the current to decrease to values smaller than those found in thermal equilibrium. A situation must exist where photoexcitation actually

Figure 5.7. Photocurrent variation with time during photoexcitation by a primary light source with wavelength of 535 nm, and secondary light sources with wavelengths between 650 and 800 nm, showing the phenomenon of optical quenching of photoconductivity.

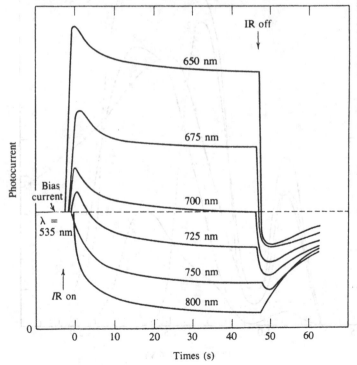

allows free carriers to be removed from the band and localized at suitable imperfections in steady state.

### Photoconductivity saturation with illumination intensity

In a material in which high photosensitivity exists because of imperfection sensitization, an increase in the photoexcitation intensity above a certain critical limit may result in an effective saturation of the photoconductivity in a condition in which $\tau \propto G^{-1}$. The effect can be correlated with the situation where the imperfections responsible for

Figure 5.8. Optical quenching spectra for a variety of CdS crystals. (1) and (2) Photosensitive crystals. (3) Insensitive crystal. (4) Crystal with intermediate sensitivity. (5) A special crystal showing quenching to within 0.1 eV of the absorption edge. The dotted lines at higher photon energies for curves (1)–(4) indicate where the measurement of quenching is strongly affected by the high excitation caused by these same photon energies.

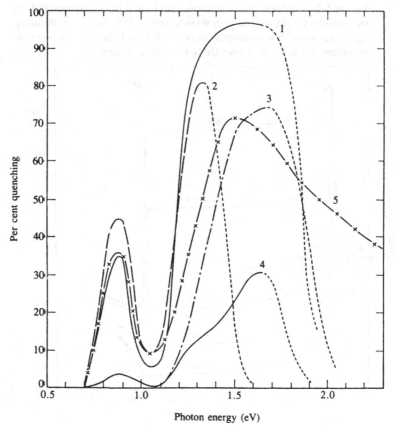

Per cent quenching

Photon energy (eV)

the long majority carrier lifetimes are totally occupied by minority carriers, so that subsequent excitation must produce minority carriers in sites much more favorable for recombination.

## 5.2 A qualitative two-center model

A qualitative exposition of a two-center model can be fairly simply given in terms of the energy band diagram of Figure 5.9, explicitly for the case of n-type photoconductivity. A more complete discussion of the implication of the Fermi level and demarcation level is given in Section 4.4.

Figure 5.9(*a*) shows a typical situation in an insensitive semiconductor where only recombination (here called Type I) centers are present. The relatively large value of $\beta_{nR}$ results in a small electron lifetime.

Figure 5.9. Energy band diagrams for qualitative discussion of the two-center recombination model. (*a*) Excitation and recombination processes with only recombination imperfections (here labeled Type I); (*b*) excitation and recombination processes with both recombination imperfections and sensitizing imperfections (here labeled Type II), at high temperatures and/or low photoexcitation intensities; (*c*) excitation and recombination processes with both recombination and sensitizing imperfections at low temperatures and/or high photoexcitation intensities; (*d*) optical quenching in a photosensitive material with both recombination and sensitizing imperfections present.

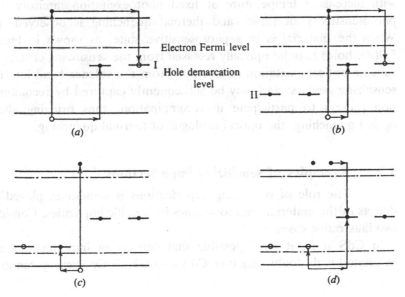

Figure 5.9(b) shows the situation where both recombination and sensitizing (here called Type II) centers are present, but the temperature is high enough and/or the photoexcitation intensity is low enough that the sensitizing centers are effectively hole traps and do not affect the recombination kinetics. The sensitivity is essentially the same as in the absence of the sensitizing centers in (a). Figure 5.9(c) shows the situation at low temperatures and/or high photoexcitation intensities, where the sensitizing centers behave like recombination centers. The electrons present in the sensitizing centers in (b) are now effectively transferred to the recombination centers (if the density of recombination and sensitizing centers is much larger than the density of free carriers), thereby decreasing the density of empty recombination centers, and the electron lifetime increases because of the small value of $\beta_{nS}$.

The major phenomena we have described above may be described in this context. (1) The addition of the sensitizing imperfections, and operation of the material under conditions such that these imperfections behave like low-recombination coefficient recombination centers and not simply like hole traps, leads to an increase in the electron lifetime. (2) When the sensitizing centers change their behavior from hole traps to recombination centers with increasing photoexcitation intensity at fixed temperature, the photosensitivity increases and supralinear photoconductivity is observed. (3) When the sensitizing centers change their behavior from recombination centers to hole traps with increasing temperature at fixed photoexcitation intensity, the photosensitivity decreases and thermal quenching is observed. (4) When the material is in a photosensitive state, as shown in Figure 5.9(d), holes may be optically released from the sensitizing centers by suitable optical excitation of electrons from the valence band into the sensitizing centers, and may be subsequently captured by recombination centers to participate in recombination, thus bringing about optical quenching, the optical analogue of thermal quenching.

## 5.3    Examples of sensitizing imperfections

The role of sensitizing imperfections is sometimes played by defects in the material, and sometimes by specific impurities. Consider two illustrative cases.

In CdS it is at least possible that sensitizing imperfections are associated with doubly negative Cd vacancies. A Cd vacancy can exist in a singly negative state describable as

$$V_{Cd}' + h \rightarrow V_{Cd}^x \tag{5.1}$$

$$V_{Cd}^x + e \rightarrow V_{Cd}' \tag{5.2}$$

where we use the nomenclature of (') to mean a negative charge, and (ˣ) to mean a zero charge with respect to the normal site in the perfect crystal. The hole capture coefficient corresponding to Eq. (5.1), $\beta_p'$ is for a Coulomb attractive process, which is larger than the electron capture coefficient corresponding to Eq. (5.2), $\beta_p' > \beta_n^x$. The Cd vacancy can also capture a second electron, however, describable as

$$V_{Cd}'' + h \rightarrow V_{Cd}' \tag{5.3}$$

$$V_{Cd}' + e \rightarrow V_{Cd}'' \tag{5.4}$$

Here the capture coefficient for the hole in Eq. (5.3), $\beta_p''$, is again for a Coulomb attractive process, which is much larger than the electron capture coefficient corresponding to Eq. (5.4), $\beta_p'' \gg \beta_n'$.

A typical example of a sensitizing center being associated with an impurity is the case of Zn impurity in Si. Zn substituted on a Si site is a double acceptor and can be described as

$$Zn'' + h \rightarrow Zn' \tag{5.5}$$

$$Zn' + e \rightarrow Zn'' \tag{5.6}$$

Once again the electron capture coefficient of Eq. (5.6) is much smaller than the hole capture coefficient of Eq. (5.5).

These examples illustrate the common experience that the incorporation of sensitizing centers with their strongly asymmetric capture coefficients increases the majority carrier lifetime but decreases the minority carrier lifetime.

Table 5.1 summarizes some of the available data on sensitizing centers in typical materials. On a purely empirical basis one may note that the minority carrier ionization energy of these sensitizing centers is about one-third of the band gap.

## 5.4  Fermi level and demarcation level analysis

Supralinear photoconductivity and thermal quenching of photoconductivity data can be analyzed using the steady state Fermi level and demarcation level analysis described in Eq. (2.84), relating the hole demarcation level of the sensitizing centers and the steady state

Table 5.1. *Data on sensitizing centers in typical materials*

| Material | Bandgap (eV) | Hole ionization energy of sensitizing center (eV) |
|---|---|---|
| CdS | 2.4 | 1.2 |
| CdSe | 1.7 | 0.6 |
| ZnS | 3.7 | 1.2 |
| ZnSe | 2.6 | 0.6 |
| GaAs | 1.4 | 0.4 |
| GaP | 2.2 | 0.7 |
| InP | 1.2 | 0.4 |
| Ge:Mn | 0.7 | 0.3 |
| Si:Zn | 1.1 | 0.6 |
| HgI$_2$ | 2.1 | 0.5 |
| ZnTe | 2.1 | 0.3[a] |
| GaSe | 2.0 | 0.5[a] |

[a] Possible cases of p-type photoconductivity, in which case the quoted values are for electron ionization energies.

electron Fermi level:

$$E_{Dp} = (E_c - E_{Fn}) + kT \ln[(m_h^*/m_e^*)(S_p/S_n)] \qquad (5.7)$$

if we choose to set $E_v = 0$.

Consider the physical meaning of the 'breakpoints' in plots of photoconductivity vs photoexcitation intensity at fixed temperature, and plots of conductivity vs temperature at fixed photoexcitation intensity. The photosensitivity starts to decrease when the sensitizing center begins to be transformed from a recombination center to a hole trap (for n-type photoconductivity). Therefore we may conclude that the break from high sensitivity (the 'upper breakpoint') occurs when the hole demarcation level for the sensitizing centers lies at the level of the sensitizing imperfections.

Therefore from Eq. (5.7),

$$E_S = (E_c - E_{Fn})_{ub} + kT_{ub} \ln[(m_h^*/m_e^*)(S_p/S_n)] \qquad (5.8)$$

where $E_S$ is the hole ionization energy of the sensitizing imperfections, and the subscript 'ub' signifies the value at the 'upper breakpoint.' Alternatively this relationship can be written in terms of the electron density at the upper breakpoint, $n_{ub}$.

$$\ln n_{ub} = -E_S/kT_{ub} + \ln[N_{v(ub)}(m_e^*/m_h^*)^{1/2}(S_p/S_n)] \qquad (5.9)$$

since $(E_c - E_{Fn})_{ub} = kT_{ub} \ln(N_c/n_{ub})$ and $N_c/N_v = (m_e^*/m_h^*)^{3/2}$, and where $N_{v(ub)} = N_{v(300 \text{ K})}(T_{ub}/300)^{3/2}$.

It is therefore possible to make two alternative and equivalent plots. In the first, one follows Eq. (5.8), plots $(E_c - E_{Fn})_{ub}$ vs $T_{ub}$ and obtains a straight line with energy intercept of $E_S$ and slope of $-k \ln[(m_h^*/m_e^*)(S_p/S_n)]$. In the second, one follows Eq. (5.9), plots $\ln(n_{ub} T_{ub}^{-3/2})$ vs $1/T_{ub}$ and obtains a straight line with slope of $-E_S/k$ and vertical axis intercept of $\ln[N_{v0}(m_e^*/m_h^*)(S_p/S_n)]$. Although not accurately temperature-corrected, the plot for the upper breakpoint in Figure 5.5 shows the general behavior predicted by Eq. (5.9), indicating that the activation energy for that data gives an approximate value for $E_S$. We will say more about the interpretation of Figure 5.5 in a following section. Note that such analysis of supralinear photoconductivity or thermal quenching of photoconductivity yields values of sensitizing center ionization energy $E_S$ and the ratio of its capture cross sections $(S_p/S_n)$, but not the specific values of the cross sections themselves.

A similar analysis can be applied to data of optical quenching of photoconductivity. From the low photon energy cutoff of the spectral response of optical quenching, it is possible to determine the value of $E_S$. From the conditions for the onset of optical quenching with varying intensities of quenching light, it is possible to determine the majority carrier capture cross section by itself. By carrying out measurements as a function of temperature, it is then possible to determine the temperature dependence of the majority carrier capture cross section.

Figure 5.10 shows optical quenching data for a particular CdS crystal at room temperature for different intensities of the quenching light. The onset of optical quenching occurs for higher photocurrents (higher 2.20 eV photon fluxes) for higher intensities of the quenching 1.30 eV light. By analogy with the case of thermal quenching, we may propose that optical quenching occurs when the rate of electron excitation to empty sensitizing levels from the valence band $G_Q$ is just equal to the rate of electron capture by the sensitizing levels.

If this onset of quenching occurs for an electron density $n_b$, then

$$n_b(N_S - n_S)_b S_n v_n = G_Q = F_Q S_{opt}(N_S - n_s)_b \tag{5.10}$$

where $S_{opt}$ is the optical cross section for the optical transition of electrons from the valence band to the sensitizing level, the corresponding absorption coefficient being $\alpha = S_{opt}(N_S - n_s)_b$, and $F_Q$ is the photon flux per unit area per second of quenching radiation. The electron capture cross section is therefore given by

$$S_n = F_Q S_{opt} / n_b v_n \tag{5.11}$$

Figure 5.11 shows similar optical quenching data for another crystal of CdS for a single quenching intensity at two different temperatures. The fact that approximately the same value of $n_b$ is found for the breakpoint at both temperatures, indicates that, except for the relatively small temperature variation of $v_n \propto T^{1/2}$, the electron capture cross section for the sensitizing centers in this crystal is relatively temperature independent. Electron capture cross sections have been measured in this way for a variety of materials showing two-center photosensitivity effects with the results summarized in Table 5.2. The

Figure 5.10. Variation of photocurrent excited by 2.20 eV photons when exposed to varying intensities of 1.30 eV photons in CdS at room temperature.[4] The breakpoint for the onset of quenching can be used to calculate the electron capture cross section.

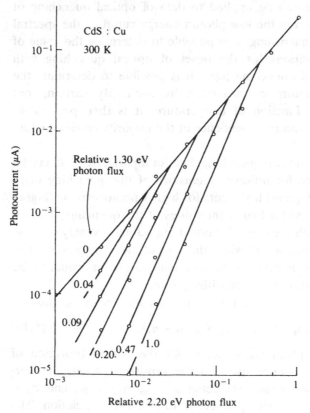

values of the electron cross section for the different materials have about the same order of magnitude, lying between $10^{-22}$ and $10^{-20}$ cm$^2$, and in general showing little temperature dependence. The agreement may be considered acceptable when uncertainties in the values of $S_{opt}$ are taken into account.

## 5.5 Negative photoconductivity

Photoexciting a semiconductor may cause negative photoconductivity, i.e., a decrease in the steady state conductivity to a value lower than that of the thermal equilibrium dark conductivity, provided that several conditions are met. Figure 5.12 illustrates the situation.

Figure 5.11. Variation of photocurrent in CdS excited by 2.41 eV photons when exposed to the same intensity of quenching radiation at 90 K and 295 K.[4]

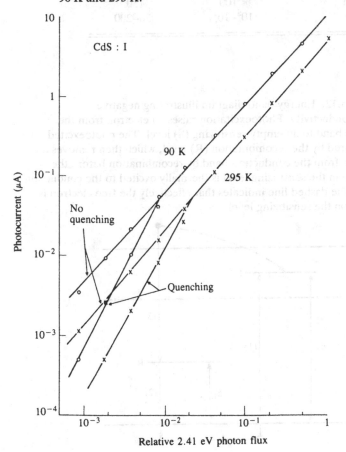

Relative 2.41 eV photon flux

Table 5.2. *Values of electron capture cross section of*
*sensitizing centers*

| Material | Electron cross section $(10^{-22}\ cm^2)$ | Temperature range (K) |
|---|---|---|
| CdS:I | $10 \pm 3$ | 90–350 |
| CdS (undoped) | $1.1 \pm 0.1$ | 300 |
| CdS:Cu | $76 \pm 44$ | 238–326 |
| CdS:Ag | $62 \pm 25$ | 81–203 |
| $CdS_{0.25}Se_{0.75}$ | 3 | 83 |
| CdSe:Cl:Cu | 2 | 83 |
| GaAs:Si:Cu | $6 \pm 2$ | 81–166 |
| GaAs:Si:Cu | $3 \pm 1$ | 81–141 |
| InP:Cu | $36 \pm 5$ | 81–141 |
| ZnS | 0.1 | 300 |
| Ge:Mn | $<100$ | 150 |
| Si:Zn | $10^2$–$10^3$ | 30–200 |

Figure 5.12. Energy band diagram illustrating negative
photoconductivity. Photoexcitation raises an electron from the
valence band to an empty sensitizing (S) level. The photoexcited hole
is captured by the recombination (R) level, which then removes an
electron from the conduction band by recombination before the
electron in the sensitizing level is thermally excited to the conduction
band. The dashed line indicates that effectively the free electron is
placed on the sensitizing level.

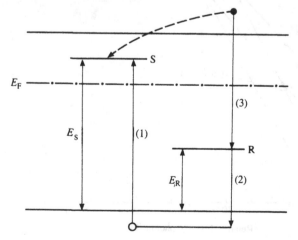

Photoexcitation effectively transfers an electron from the conduction band to a sensitizing center that was empty in the dark.

Three conditions must all be satisfied:

(1) The levels of the sensitizing imperfections must lie above the dark Fermi level, so that they are empty in the dark, i.e., $E_S > E_F$.

(2) The recombination imperfections must compete favorably with the sensitizing imperfections for photoexcited holes, so that recombination can proceed through the recombination imperfections, i.e. $\beta_{pR} \approx \beta_{pS}$.

(3) The rate of thermal reexcitation of electrons from the sensitizing levels to the conduction band must be much less than the rate of capture of electrons from the conduction band by the recombination imperfections, i.e.,

$$n_S N_c \beta_{nS} \exp[-(E_c - E_S)/kT] \ll n(N_R - n_R)\beta_{nR} \quad (5.12)$$

Since $n_S = (N_R - n_R)$ for particle conservation, and $n/N_c = \exp[-(E_c - E_F)/kT]$, the necessary condition is

$$S_{nS}/S_{nR} \ll \exp[-(E_S - E_F)/kT] \quad (5.13)$$

If a typical value for the cross section on the ratio on the left of Eq. (5.13) is about $10^{-6}$, then the requirement is that $(E_S - E_F) < 0.34$ eV.

## 5.6 Photoconductivity saturation with increasing photoexcitation rate

It is observed that when the photoexcitation rate is increased beyond a critical value in a photosensitized semiconductor, the photoconductivity effectively saturates. Experimental data for a CdSSe crystal at several different temperatures is given in Figure 5.13.[5] Saturation of the photoconductivity occurs when the photoexcitation rate is sufficiently high that essentially all of the sensitizing imperfections are hole-occupied. If there were only free electrons and sensitizing centers involved, the saturated value of the free electron density would simply be equal to the sensitizing center density, thus providing a method for determining the density of the sensitizing imperfections. In the real case, there are also electrons in traps, so that $n_{sat} + n_t = N_S$, where $n_t$ is the density of light-filled traps at the saturation of photoconductivity. It follows that

$$n_{sat} = N_S - N_t/[\exp(E_t - E_F)/kT + 1] \quad (5.14)$$

Figure 5.13. Photocurrent as a function of photon flux incident on a crystal of CdSSe at different temperatures, exhibiting saturation of photoconductivity.[5]

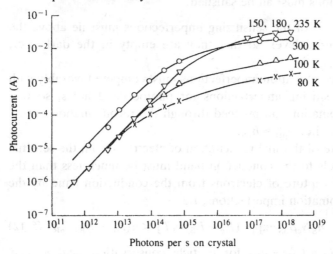

Figure 5.14. The solid curve shows the variation of the density of free electrons at saturation as a function of the position of the steady state Fermi level at saturation. The dashed curve is the sum of the density of free electrons at saturation and the calculated density of electrons trapped in a trap with $E_c - E_t = 0.051$ eV, and $N_t = 2 \times 10^{16}$ cm$^{-3}$.[5]

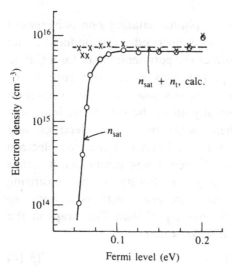

where $N_t$ is the total trap density and $E_t$ is the height of the electron trap level above the valence band.

The solid curve of Figure 5.14 shows the variation of the density of free electrons at saturation as a function of the location of the steady state Fermi level at saturation. From the temperature-independent portion of the curve, it may be concluded that $N_S = 7 \times 10^{15}\,\text{cm}^{-3}$ for this sample. The whole curve is fit well by Eq. (5.14) if values of $N_t = 2 \times 10^{16}\,\text{cm}^{-3}$ and $E_c - E_t = 0.051\,\text{eV}$ are chosen.

## 5.7 A simple analytical two-center model

Figure 5.15 shows an energy band diagram for a simple two-center recombination model with sensitizing centers lying $E_S$ above the valence band, recombination centers lying $E_R$ above the valence band, and electron traps lying $E_t$ above the valence band.[6] To simplify the notation all recombination coefficients are assumed equal to $\beta$, except for electron capture by sensitizing imperfections, for which $\beta' \ll \beta$. We neglect thermal emptying completely for the recombination centers, as well as thermal excitation of electrons from the sensitizing centers to the conduction band, and thermal excitation of holes from traps to the valence band. For convenience we define the

Figure 5.15. Energy band diagram for a simple two-center recombination model with sensitizing centers lying $E_S$ above the valence band, recombination centers lying $E_R$ above the valence band, and electron traps lying $E_t$ above the valence band. To simplify the notation all recombination coefficients are assumed equal to $\beta$, except for electron capture by sensitizing imperfections, for which $\beta' \ll \beta$.

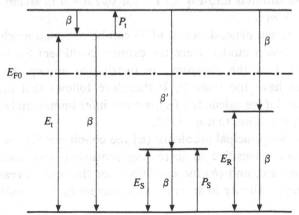

thermal excitation terms, $P_t = N_c \beta \exp[-(E_c - E_t)/kT]$ and $P_S = N_v \beta \exp(-E_S/kT)$.

Particle conservation requires

$$n + n_t = p + (N_R - n_R) + (N_S - n_S) \tag{5.15}$$

The continuity equations are

$$dn/dt = G - n\beta(N_R - n_R) - n\beta(N_t - n_t) - n\beta'(N_S - n_S) + n_t P_t \tag{5.16}$$

$$dp/dt = G - p\beta(n_S + n_R + n_t) + (N_S - n_S)P_S \tag{5.17}$$

$$dn_R/dt = n\beta(N_R - n_R) - n_R\beta p \tag{5.18a}$$

so that

$$n_R = [n/(n + p)]N_R \tag{5.18b}$$

$$dn_S/dt = n\beta'(N_S - n_S) - n_S\beta_p + (N_S - n_S)P_S \tag{5.19a}$$

so that

$$n_S = \{(\beta'n + P_S)/[\beta p + (\beta'n + P_S)]\}N_S \tag{5.19b}$$

$$dn_t/dt = n\beta(N_t - n_t) - n_t\beta p - n_t P_t \tag{5.20a}$$

so that

$$n_t = \{\beta n/[\beta(n + p) + P_t]\}N_t \tag{5.20b}$$

To obtain a general solution it is necessary to substitute Eqs. (5.18b), (5.19b) and (5.20b) into Eq. (5.15) in order to solve for $p(n)$. Then if this result is substituted into Eq. (5.17) for $dp/dt = 0$ in steady state, we can calculate $n(G)$.

But the functional dependence of $n(G)$ can be obtained much more simply in the chosen model. Here the capture coefficient for holes is independent of how the electrons are distributed among the three levels since all have the same $\beta$. It therefore follows that approximately $p \propto G$ and if we calculate $n(p)$, we can infer immediately $n(G)$, e.g., if $p \propto n^m$, it follows that $n \propto G^{1/m}$.

We address two principal problems: (a) the conditions for the lower breakpoint from insensitive to increasing sensitivity with increasing photoexcitation rate, and (b) the conditions for the upper breakpoint from increasing sensitivity to sensitive with increasing photoexcitation rate.

*Low photoexcitation rate: beginning of supralinear
photoconductivity*

The relevant conditions are that ($a$) the hole Fermi level lies above the sensitizing levels: $\beta p \ll P_S$, ($b$) the hole demarcation level for the sensitizing levels lies above these levels: $\beta'n \ll P_S$, ($c$) $p \ll n$, ($d$) $n \ll N_t$. Substitution of Eqs. (5.18b), (5.19b) and (5.20b) into Eq. (5.15) subject to these inequalities gives

$$p = [n^2 P_S (\beta N_t + P_t)]/\{[n(P_S + \beta N_S) + P_S N_R](\beta n + P_t)\} \quad (5.21)$$

Four possible ranges may be defined:

(1) $n(P_S + \beta N_S) \ll P_S N_R$ and $\beta_n \ll P_t$. The second condition implies that the traps are behaving like traps and are primarily empty. Under these conditions $p \propto n^2$ and therefore $n \propto G^{1/2}$.

(2) $n(P_S + \beta N_S) \ll P_S N_R$ and $\beta_n \gg P_t$. Here the second condition implies that the traps are behaving like recombination centers and are primarily electron occupied. Under these conditions $p \propto n$ and $n \propto G$.

(3) $n(P_S + \beta N_S) \gg P_S N_R$ and $\beta_n \ll P_t$. Now $p \propto n$ and $n \propto G$.

(4) $n(P_S + \beta N_S) \gg P_S N_R$ and $\beta_n \gg P_t$. For this condition $p$ is a constant, independent of $G$, which implies that $n \propto G^{>1}$.

There are therefore two conditions for the onset of supralinearity at the lower breakpoint.

If the traps are filled, $\beta n \gg P_t$, the criterion will be given by the transition from range (2) to range (4) above: $n(\beta N_S + P_S) = P_S N_R$. Under the condition that $\beta N_S \gg P_S$, this means that supralinear photoconductivity starts when

$$\beta n_{lb} = (N_R/N_S)P_S \quad (5.22)$$

where $n_{lb}$ is the electron density corresponding to the lower breakpoint. A plot of $\ln n_{lb}$ vs $1/T_{lb}$, where $T_{lb}$ is the temperature for the lower breakpoint, has an activation energy corresponding to $E_S$, consistent with the results shown in Figure 5.5.

Another possibility for the onset of supralinear photoconductivity corresponds to the case where the transition occurs from range (3) to range (4). It occurs when $n(P_S + \beta N_S) \gg P_S N_R$ and occurs when the traps fill corresponding to $\beta n_{lb} = P_t$. In this case a plot of $\ln n_{lb}$ vs $1/T_{lb}$ has an activation energy corresponding to $(E_c - E_t)$.

*High photoexcitation rate: end of supralinear photoconductivity*

The following relationships hold in this case: (a) $\beta n$, $\beta p \gg P_t$, (b) $\beta p \ll (\beta'n + P_S)$, (c) $p \ll n$, (d) $n \ll N_t$, $N_R$. Substitution of Eqs. (5.18b), (5.19b) and (5.20b) into Eq. (5.15) subject to these inequalities gives

$$p = [nN_T(\beta'n + P_S)]/[n\beta N_S + N_R(\beta'n + P_S)] \tag{5.23}$$

There is only one condition corresponding to the end of supralinearity at the upper breakpoint: $\beta'n = P_S$. When $\beta'n \ll P_S$, $p = $ constant and $n \propto G^{>1}$ (since $\beta n \gg (N_R/N_S)P_S$ in this range), whereas when $\beta'n \gg P_S$, $p \propto n$ and $n \propto G$. The upper breakpoint therefore corresponds to $\beta'n = P_S$, and a plot of $\ln n_{ub}$ vs $1/T_{ub}$ has an activation energy of $E_S$.

This analysis fits well with the data of Figure 5.5. The upper breakpoint curve corresponds to $\beta'n_{ub} = P_S$ and the lower breakpoint curve corresponds to $\beta n_{lb} = (N_R/N_S)P_S$. Therefore the two curves have the same activation energy, equal to $E_S$, and the ratio between the two curves at a particular temperature is given by $\beta(N_R/N_S)/\beta'$, which allows an estimate of the ratio cross sections $S_p/S_n \approx 8 \times 10^5$ for the sensitizing imperfections if $N_R \approx N_S$.

## 5.8    Calculations with the general model

More detailed calculations on a model with sensitizing centers, recombination centers, and electron traps can be made using the general model of Section 3.3.[7]

A typical example is given in Figure 5.16 which has been constructed in an effort to simulate behavior in CdSe, and shows the calculated variations of electron and hole lifetimes as a function of photoexcitation rate at several different temperatures. The results can be compared with the experimental data of Figure 5.3. For the data in Figure 5.3, the experimentally used range of photoexcitation rates corresponds to a range of $G$ from $10^{12}$ to $10^{18}$ cm$^{-3}$ s$^{-1}$. At temperatures below 210 K, no supralinear behaviour is observed experimentally over this range of $G$. Supralinearity for $G$ between $10^{13}$ and $10^{14}$ cm$^{-3}$ s$^{-1}$ is observed at 240 K, and at 315 K supralinear photoconductivity is observed over the whole range from $10^{14}$ to $10^{18}$ cm$^{-3}$ s$^{-1}$. Finally at 380 K no supralinear behavior is observed over this whole range of $G$. Comparison between Figures 5.5 and 5.16 indicates the close agreement of these experimental observations with the model predictions. The model also shows clearly that the value of the electron density, $n$, increases with $G$ until $n$ is equal to the density of the sensitizing centers; for higher $G$, $n$ stays constant $(\tau_n \approx G^{-1})$, and the electron lifetime decreases to meet the hole lifetime for very high $G$. Thus the model also shows the phenomenon of saturation of photoconductivity.

One major difference between the model predictions of Figure 5.16 and the experimental data of Figure 5.5 may be noted. In experimental measurements the range of $G$ over which the electron lifetime increases is as much as four orders of magnitude, whereas the model predicts a much more abrupt change than this. The explanation for this difference may be sought either in the fact that the actual crystal contains a distribution of traps with different depth and not just a single trap, or in material inhomogeneities not considered in the model.

The location of supralinearity on the temperature and photoexcitation rate scales is a sensitive function of the hole ionization energy of the sensitizing centers. In CdSe these centers lie 0.64 eV above the valence band, whereas in CdS they lie 1.00 eV above the valence band. Figure 5.17 compares supralinear behavior in CdS with that just discussed for CdSe, showing the strong shift to higher temperatures, consistent with the experimental observation that supralinear photoconductivity is not observed in CdS until the temperature approaches 375 K for this range of $G$.

Figure 5.16. Electron and hole lifetimes over the range exhibiting supralinear photoconductivity in a simulated CdSe case.[7] Parameters used are $m_e{}^* = 0.2m$, $m_h{}^* = 0.5m$, $N_R = N_t = 10^{15}$ cm$^{-3}$, $N_S = 10^{17}$ cm$^{-3}$, $E_R = 0.80$ eV, $E_t = 1.30$ eV, $E_S = 0.64$ eV, $E_G = 1.70$ eV, $E_F = 0.85$ eV, $\beta_{Rn} = 10^{-16}v_g$, $\beta_{Rp} = 10^{-12}(300/T)^2v_p$, $\beta_{Sn} = 10^{-21}v_n$, $\beta_{Sp} = 4 \times 10^{-12}(300/T)^2v_p$, $\beta_{tn} = 10^{-12}(300/T)^2v_n$, $\beta_{tp} = 10^{-16}v_p$ cm$^3$ s.

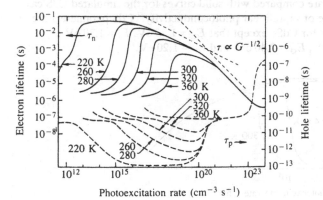

Photoexcitation rate (cm$^{-3}$ s$^{-1}$)

Of course, all quantities of interest in the model can be evaluated as a function of temperature, photoexcitation rate, imperfection densities and capture cross sections.[7] The occurrence of supralinear photoconductivity, for example, requires that $N_S > N_R$, but even only a 20% excess of $N_S$ over $N_R$ is sufficient to give definite supralinear behavior.

## 5.9     Specific examples

To conclude the discussion of this chapter we simply cite a few examples, from the very large number that might be selected, to illustrate further some of the phenomena described here.

### Identifying the sensitizing center

In many cases it has been difficult actually to identify the sensitizing center. A particular example is the role of Cu in such materials as CdS and ZnSe: does the Cu act like a sensitizing center and, if so, is Cu *the* sensitizing center commonly found? The typical optical quenching spectrum, such as is shown in Figure 5.8 has a low-energy band, apparently characteristic of a discrete level to discrete level transition and with a thermally activated step needed to produce free holes. It was suggested[8] that the sensitizing center was indeed a Cu on a Cd site, and that this low-energy quenching band was due to an internal d-shell transition in the $Cu_{Cd}{}^{\times} d^9$ configuration.

Three different crystals of CdS are compared in Figure 5.18;[4] one of them labeled CdS:I has been prepared with I donors without any Cu, a second labeled CdS:Cu has been grown with Cu impurity deliberately, and a third labeled CdS(C) does not have Cu deliberately added

Figure 5.17. The dashed curves from Figure 5.16 for the simulated CdSe case are compared with solid curves for the simulated CdS case in the range of supralinear photoconductivity.[7] CdS parameters are the same as for CdSe except that $E_R = 1.15\,\text{eV}$, $E_t = 1.70\,\text{eV}$, $E_S = 1.00\,\text{eV}$, $E_G = 2.40\,\text{eV}$, and $E_F = 1.20\,\text{eV}$.

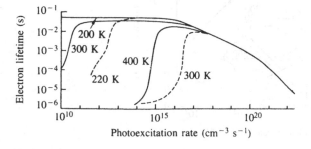

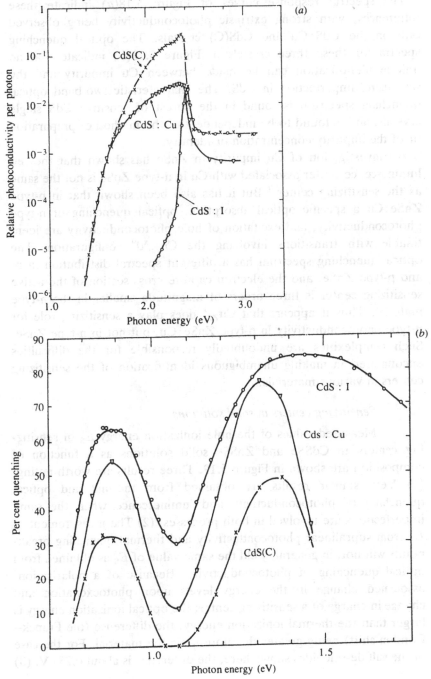

Figure 5.18.(*a*) Spectral response of photoconductivity of a high-purity CdS crystal with a trace of I (CdS : I), a crystal of CdS grown with Cu impurity (CdS : Cu), and a crystal of CdS grown at high temperatures (CdS(C)). (*b*) Optical quenching spectra for the three CdS crystals in (*a*).[4]

but was grown at a high temperature and may either have unintentional Cu from diffusion or a high density of imperfections.

The spectral response curves of Figure 5.18(a) indicate these differences, with strong extrinsic photoconductivity being observed only in the CdS:Cu and CdS(C) crystals. The optical quenching spectra for these three crystals in Figure 5.18(b) indicate that no unique identification can be made between Cu impurity and the sensitizing imperfection in CdS. The characteristic two-band optical quenching spectrum is found in the purest of sensitive CdS single crystals, and is found to be independent of the method of preparation or of the impurity concentration or identity.

An investigation of Cu impurity in ZnSe has shown that the red luminescence center associated with Cu in n-type ZnSe is not the same as the sensitizing center.[9] But it has also been shown that in p-type ZnSe:Cu a specific optical absorption, optical quenching of n-type photoconductivity, and excitation of hole photoconductivity are identifiable with transitions involving the $Cu_{Zn}{}^x d^9$ configuration. The optical quenching spectrum has a different spectral distribution in n- and p-type ZnSe, and the electron capture cross section of the active sensitizing center is three orders of magnitude smaller in the n-type material. Thus it appears that $Cu_{Zn}{}^x$ does play a sensitizing role for n-type photoconductivity in p-type ZnSe:Cu, but not in n-type ZnSe. Such complexities are undoubtedly responsible for the difficulties encountered in making unambiguous identification of the sensitizing centers in various materials.

### Sensitizing centers in solid solutions

Measured values of the hole ionization energy $E_S$ of sensitizing centers in CdSSe and ZnSSe solid solutions as a function of composition are shown in Figure 5.19. Three results are worth noting. (1) Very similar results are obtained from thermal and optical quenching of photoconductivity and luminescence when the same imperfections are involved in both processes. (2) The measurement of $E_S$ from supralinear photoconductivity and thermal quenching breakpoints will not, in general, yield the same value of $E_S$ as obtained from optical quenching of photoconductivity. Because of a polarization-associated change in the energy levels upon photoexcitation and change in charge of a sensitizing center, the optical ionization energy is larger than the thermal ionization energy, the difference (the Franck–Condon shift) being greater the more ionic the material. For the case of the sulfide-selenides shown here, the difference is about 0.25 eV. (3)

Although the bandgap varies smoothly between end members of the solid solutions, the hole ionization energy of the sensitizing centers does not. An abrupt transition from sulfide-like to selenide-like behavior occurs between 27 and 40% selenide.

### High impurity concentration effects

It is well known that imperfection ionization energies are often affected by increasing the density of impurities in the material. Observations that appear to be related to this are found in CdS crystals doped with increasing concentrations of Cu acceptor and Ga donors, using a ratio of $[Cu]/[Ga] = 1.05$ and maintaining an insulating material throughout.[12] Figure 5.20($a$) shows the variation of photocurrent with light intensity for samples with different impurity densities. As the impurity densities increase, supralinear photoconductivity regions appear much like in Figure 5.4, behaving as if the ionization energy of the sensitizing centers were decreasing with increasing impurity densities. Corresponding thermal quenching curves for the same series of samples are shown in Figure 5.20($b$), showing a decrease in the temperature for the onset of thermal quenching at higher impurity densities.

Figure 5.19. Variation of the hole ionization energy of sensitizing centers with composition in CdSSe[10] and ZnSSe[11] solid solutions. Open points are for CdSSe, solid points for ZnSSe; circles are optical ionization energies, squares are thermal ionization energies.

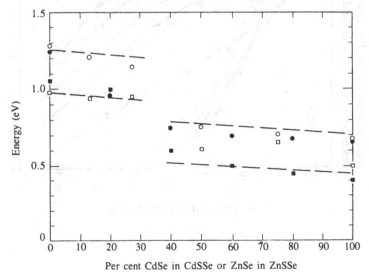

Per cent CdSe in CdSSe or ZnSe in ZnSSe

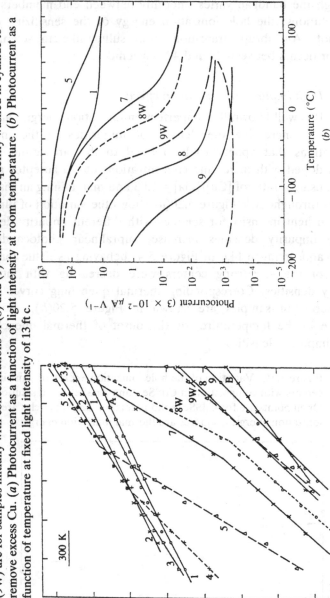

Figure 5.20. Effects of increasing Ga and Cu impurity concentrations in CdS.[12] Numbers on the curves are sample numbers and correspond to the following Cu concentrations in $10^{-5}$ moles per mole of CdS: (1) 2, (2) 4, (3) 10, (4) 20, (5) 40, (7) 200, (8) 400, and (9) 1000. In each sample the concentration of Cu is 1.05 times the concentration of Ga. Sample B is the same as (9) but without any Ga impurity. Curves labeled (8W) and (9W) are for samples initially with concentrations of (8) and (9) that were subsequently washed in cyanide to remove excess Cu. (a) Photocurrent as a function of light intensity at room temperature. (b) Photocurrent as a function of temperature at fixed light intensity of 13 ft c.

If these data are interpreted as a decrease in sensitizing center hole ionization energy, then they indicate that this ionization energy is decreased from 1.0 to 0.3 eV by high density impurity doping. It is important to note, however, that this is not an unambiguous interpretation. High impurity doping also increases the density of electron traps, and an increase in the density of traps can also cause the upper breakpoint for supralinear photoconductivity and thermal quenching of photoconductivity curves to shift in the manner measured.[13]

### Si single crystals with Zn impurity

Since Si has been for many years the prototype semiconductor, and Zn impurity provides a well-understood sensitizing imperfection in Si, it is appropriate to conclude this discussion by noting the appearance of these two-center effects in Si:Zn.[14] Zn impurity in Si has a single acceptor level Zn' lying 0.31 cV above the valence band, and a double acceptor level Zn" lying 0.57 eV above the valence band at room temperature. When the Fermi level lies above the double acceptor level, most of the Zn is in the Zn" state and behaves like a sensitizing imperfection for n-type photoconductivity. An energy band diagram for this situation is shown in Figure 5.21.

The photoconductivity excitation spectrum measured at 80 K is given in Figure 5.22($a$). A low-energy threshold at 0.52 eV is observed for photoexcitation of an electron from the Zn" level to the conduction band.

Figure 5.21. Energy band diagram for Zn impurity in Si. Labeled transitions are the following: (1) spectral response of photoconductivity threshold, (2) electron capture by Zn' center, (3) optical quenching transition.

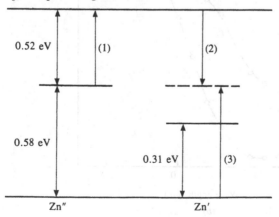

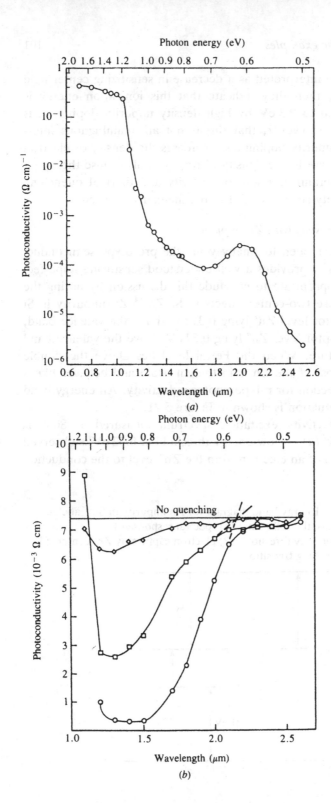

Photon energy (eV)

2.0 1.6 1.4 1.2   1.0 0.9 0.8   0.7   0.6   0.5

Wavelength (μm)

(a)

Photon energy (eV)

1.2 1.1 1.0 0.9 0.8   0.7   0.6   0.5

No quenching

Wavelength (μm)

(b)

Figure 5.22($b$) shows the optical quenching spectrum for photoconductivity in Si:Zn at 80 K for three different quenching intensities.
The low-energy threshold is 0.58 eV, the energy required to excite an
electron from the valence band to the Zn' level to form a Zn" center.
Competition between optical quenching and excitation by the same

Figure 5.22. Characteristic properties of Zn-doped Si.[14] ($a$)
Excitation spectrum for photoconductivity at 80 K. ($b$) Optical
quenching of photoconductivity at 80 K for three different quenching
intensities of (circles) $2.7 \times 10^{14}$, (squares) $3.8 \times 10^{13}$, and (diamonds)
$4.3 \times 10^{12}$ cm$^{-2}$ s$^{-1}$. ($c$) Electron density (reciprocal of the Hall
coefficient times the electron charge) as a function of photoexcitation
rate showing photoconductivity saturation for 0.9 $\mu$m (circles) and
2.3 $\mu$m (diamonds) excitation. (d) Electron capture cross section of
Zn' centers determined from onset of optical quenching for both
0.9 $\mu$m (circles) and 2.3 $\mu$m (squares) primary photoexcitation.

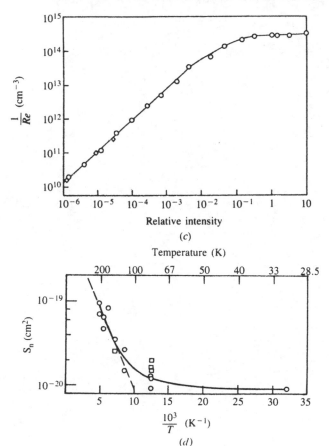

photons causes the minimum of photoconductivity seen in Figure 5.22($a$) in the vicinity of 1.7 $\mu$m (0.73 eV).

Variation of the intensity of photoexcitation at low temperatures would be expected to yield two results characteristic of sensitizing centers. ($a$) Changing the charge of the Zn impurities from $-2$ to $-1$ as the result of hole capture would be expected to yield an increased electron mobility. Actually an increase in electron mobility from 4300 to about 7000 cm$^2$ V$^{-1}$ s$^{-1}$ is observed over the intensity range shown. ($b$) If sufficiently high intensities are used to empty all the Zn″ centers, a saturation of photoconductivity would be expected as shown in Figure 5.22($c$).

Finally the electron capture cross section corresponding to Zn′ + e′ → Zn″ can be determined from the onset of optical quenching of photoconductivity. Values given in Figure 5.22($d$) are assuming an optical cross section for the quenching transition of 2 × 10$^{-15}$ cm$^2$.

# 6
# Recombination processes

Any discussion of electron capture and recombination depends critically in a quantitative sense on the magnitudes and temperature dependences of the capture cross sections or coefficients involved. In this chapter we take a brief look at some of the fairly simple ways of understanding the magnitudes of these significant quantities, recognizing that a detailed theoretical prediction of them is still often an area of active research.

When an electron or hole has been photoexcited to a higher energy state, there are three fundamental processes by which the energy can be given up when the carrier returns to its thermal equilibrium state. This excess energy can be given up by (1) emitting photons – a radiative recombination, commonly called luminescence; (2) emitting phonons either simultaneously or sequentially, a non-radiative process; or (3) raising other free carriers, either electrons or holes, to higher energy states in their bands, another non-radiative process called Auger recombination, after which they may relax by the emission of phonons.

## 6.1 Radiative recombination

### Intuitive approach

A semiqualitative, intuitive approach to the capture cross section associated with a radiative recombination is based upon considering a free electron interacting with a recombination center. The free electron is travelling with a thermal velocity of about $10^7 \, \mathrm{cm \, s^{-1}}$, and the ground state wavefunction of the recombination center extends a distance of the order of $10^{-7} \, \mathrm{cm}$, giving a passage time of about $10^{-14} \, \mathrm{s}$. Since the time required for an atomic radiative recombination is of the order of $10^{-8}$ s, one pass in $10^6$ yields radiative recombination on the average. If the geometric cross section of an imperfection is of the order of $10^{-15}$ cm$^2$, this means that the radiative recombination cross section is of the order of $10^{-21}$ cm$^2$. However limited this intuitive approach may be, it is helpful in indicating that the cross section for radiative recombination is likely to be considerably smaller than a geometrical cross section.

### Detailed balance for intrinsic recombination

Here we attempt to approximate the capture cross section appropriate to the radiative recombination of free electrons with free holes across the bandgap of the semiconductor. Consider an intrinsic semiconductor in thermal equilibrium with its surroundings. Blackbody radiation falls on the semiconductor at the rate of

$$(2\pi v^2/c^2)\{1/[\exp(hv/kT) - 1]\}\Delta v \text{ photons cm}^2 \text{ s}^{-1} \qquad (6.1)$$

Suppose that these photons are absorbed in a thickness $d$. Then if we equate the excitation rate and the recombination rate we have

$$(2\pi v^2/c^2 d)\{1/[\exp(hv/kT) - 1]\}\Delta v = n_i^2 v_n S_{cv}(1/2r^2) \qquad (6.2)$$

where the final term on the right is an optical term with a factor of 2 since both directions are involved, and the index of refraction $r$ is involved because of the reduced fraction of emission due to total reflection. If now we take $hv = E_G$, $h\Delta v = kT$, and $d = 1/\alpha$ where $\alpha$ is the optical absorption constant for intrinsic absorption near $E_G$, Eq. (6.2) becomes

$$(2\pi E_G^2/h^3 c^2 d)\{kT/[\exp(E_G/kT) - 1]\} = N_c N_v \exp(-E_G/kT)v_n S_{cv}/2r^2 \qquad (6.3)$$

and the capture cross section is given approximately by

$$S_{cv} \approx 5 \times 10^{-25}(r^2/d)E_G^2(m_e^* m_h^*)^{-3/2}(300/T)^{5/2} \text{ cm}^2 \qquad (6.4)$$

Substituting appropriate quantities for a range of semiconductors including Ge, Si, GaAs, GaP, CdS and InSb, we find that $S_{cv}$ is predicted to vary from $1.8 \times 10^{-19}$ cm$^2$ (Si) to $9.3 \times 10^{-18}$ cm$^2$ (GaP).

### Dipole approximation

A third simple approximation for the magnitude of the radiative capture cross section may be made by considering the matrix element for a spontaneous radiative transition between a completely free electron state ($\exp(i\mathbf{k} \cdot \mathbf{r})$), neglecting the Bloch modulation factor, and a 1s hydrogenic ground state $(\exp(-r/a_0))$ in the dipole approximation.[1] The resulting radiative cross section is about $10^{-21}$–$10^{-22}$ cm$^2$.

We conclude that the capture cross section for radiative recombination is considerably smaller than a geometric cross section, and depending on the material and the exact transitions involved, is expected to lie between $10^{-18}$ and $10^{-22}$ cm$^2$. More detailed analysis is required to estimate the temperature dependence of the cross section, but it appears that the temperature dependence would be small for a neutral center, and much stronger for a repulsive center.

The type of radiative recombination described above is only one of several different ways of achieving radiative recombination. A simple schematic summary of the three main ways is given in Figure 6.1. Figure 6.1($a$) pictures the recombination between a free carrier and a localized carrier to which the above discussion has been directed. If a sufficiently high density of both donors, with ionization energy of $E_D$, and acceptors, with ionization energy of $E_A$, is present so that the wavefunction of electrons localized on the donors overlaps with the wavefunction of holes localized on the acceptors, it is possible for a transition to occur between the localized electron and hole, as shown in Figure 6.1($b$). Such a radiative recombination is often called a 'pair transition' and is characterized by (i) an emission energy given by

$$\hbar\omega_{pt} = (E_G - E_D - E_A) + q^2/(4\pi\epsilon_r\epsilon_0 R) \qquad (6.5)$$

where $R$ is the distance between the participating donor and acceptor, and (ii) a peak emission energy that shifts with photoexcitation intensity, or time during decay after emission, due to the change in

Figure 6.1. Three major types of radiative recombination: ($a$) recombination between a free carrier and a localized carrier, ($b$) pair recombination between a nearby donor and acceptor, and ($c$) recombination between the atomic levels of an impurity ion.

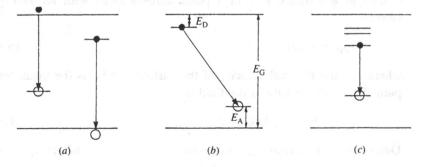

($a$)          ($b$)          ($c$)

energy with $R$ given in Eq. (6.5) as well as the dependence of recombination probability on $R$. Finally Figure 6.1($c$) shows the third possibility of a radiative recombination occurring between the excited state and the ground state of an impurity in the semiconductor with incomplete atomic levels; a typical impurity is Mn that behaves as a luminescence center in many different semiconductors.

## 6.2    Phonon emission: Coulomb attractive imperfection

### Simplest model

The simplest model for a Coulomb attractive center is to set the Coulomb potential energy equal to the average electron energy for a critical distance $r_c$:

$$q^2/4\pi\epsilon_r\epsilon_0 r_c = kT \tag{6.6}$$

If the distance from the imperfection $r < r_c$, then the electron is captured. No detailed consideration of the mechanism of energy dissipation upon capture is included in this simple form of the model. The capture cross section is given by

$$S = \pi r_c^2 = q^4/16\pi\epsilon_r^2\epsilon_0^2 k^2 T^2 \tag{6.7}$$

This cross section is of the order of $10^{-12}(300/T)\,\mathrm{cm}^2$, and therefore even at room temperature is considerably larger than a geometric cross section.

### Drift equal to diffusion

A second approach to the cross section for a Coulomb attractive center is to choose the critical radius $r_c$ such that the probability of an electron drifting into the center is equal to the probability of an electron diffusing away. A particle in Brownian motion at a distance $r$ from a point diffuses away with an average velocity

$$v_{\mathrm{diff}} = v(2d/3r) \tag{6.8}$$

where $v$ is the thermal velocity of the particle, and $d$ is the mean free path. The drift velocity in the field is

$$v_d = \mu E = \mu q/4\pi\epsilon_r\epsilon_0 r_c^2 \tag{6.9}$$

Determine $r_c$ by setting $v_{\mathrm{diff}} = v_d$, using $\mu = (q/m_e^*)\tau_{\mathrm{sc}}$ where $\tau_{\mathrm{sc}}$ is the scattering relaxation time, $v = d/\tau_{sc}$, and $m_e^* v^2/2 = kT$. Finally calculate $S = \pi r_c^2$:

$$S = \tfrac{9}{16} q^4 / 16\pi \epsilon_r^2 \epsilon_0^2 k^2 T^2 \qquad (6.10)$$

which differs from Eq. (6.7) only by the multiplying factor of $(\tfrac{9}{16})$.

### Cascade capture with multiphonon emission

Since the maximum energy of a single phonon is usually no greater than 30–50 meV, the dissipation of an energy of the order of 1 eV in recombination requires 20–30 phonons to be emitted. (Note that this is quite different than for a radiative transition, where an energy of the order of 1 eV can be dissipated by the emission of a single photon.) If these phonons are required to be emitted simultaneously in a single discrete capture transition, the corresponding capture cross section is fairly small. A theoretical calculation[2] indicated a cross section of $10^{-16}$ cm$^2$, and measured values were appreciably smaller, $S \approx 10^{-19}$ cm$^2$.

Nevertheless the simple expectation that a Coulomb attractive center should have a capture cross section of the order of $10^{-12}$ cm$^2$ at room temperature is substantiated by experiment. The reconciliation between this result and the low-probability of a simultaneous multiphonon recombination process was made with a recognition of the sequence of excited states in a Coulomb attractive center.[3] The existence of these excited states makes it possible for an electron to be captured in a high-lying state with the emission of only a few phonons. In time the 'captured' electron can then drop down from one excited state to the next lower state with the emission of only a few phonons, and with greatly diminished probability of escape before dropping down to the next lower state. Therefore it is possible for the many phonons to be emitted sequentially rather than simultaneously, and the effective cross section reflects the large $1/r$ dependence of the Coulomb attractive center.

Another dramatic multiphonon recombination process has been described for specific deep imperfection states in semiconductors.[4] A schematic diagram is shown in Figure 6.2. Non-radiative capture of an electron is accompanied by the emission of many phonons corresponding to a major change in the position of the energy level before and after capture. The capture cross section varies as $S \approx S_\infty \exp(-E_\infty/kT)$ with values of $S_\infty$ between $10^{-14}$ and $10^{-15}$ cm$^2$, like a geometric cross section, and with values of $E_\infty$ between 0 and 0.56 eV. Actual results for a defect in n-type GaAs are given in Figure 6.3, showing a nearly exponential temperature variation of the cross section with values between $5 \times 10^{-21}$ cm$^2$ below 200 K and $2 \times 10^{-18}$ cm$^2$ at 400 K.

## 6.3    Coulomb repulsive imperfection

When the sign of the charge on the carrier is the same as that on the capturing imperfection, a repulsive Coulomb potential exists between them that greatly reduces the capture cross section, whether by emission of a photon or of phonons. If thermal excitation of the carrier over the repulsive barrier is required, the cross section has the form $S = S_0 \exp(-\Delta E/kT)$, where $S_0$ is a geometric cross section and $\Delta E$ is the height of the repulsive barrier. For a typical value of $\Delta E = 0.4$ eV at 300 °K, $S \approx 10^{-6} S_0 = 10^{-21}$ cm$^2$, and a strong temperature dependence of the cross section would be expected. An alternative is that capture proceeds by thermally assisted tunneling through the repulsive barrier, in which case $S = S_0 T(E') \exp(-E'/kT)$, where $T(E')$ is the tunneling probability at an energy $E'$, producing a slightly larger cross section with a smaller temperature dependence.

Figure 6.2. Diagram illustrating how non-radiative capture with emission of many phonons occurs.[4] The equilibrium positions of the lattice coordinate and the energy level, before and after capture, are indicated by the dashed lines. The shaded regions within the energy gap indicate how the energy of the level changes as the lattice vibrates. The smaller arrows represent the amplitudes of the thermal vibrations, before and after capture of an electron. The large arrow represents the amplitude of the lattice vibrations about the new equilibrium position, immediately after capture.

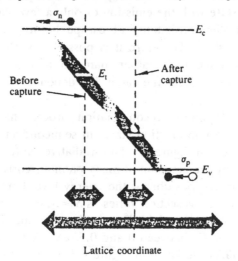

Lattice coordinate

## 6.4 Non-radiative recombination: Auger

In the Auger recombination process, the energy released upon recombination is given to another free carrier. In principle the other free carrier could be either a majority or a minority carrier, but since the probability of the energy transfer is proportional to the density of the other free carriers, a larger effect is expected for majority carriers. The rate of Auger-controlled recombination for electrons in an n-type semiconductor with $N_I$ recombination imperfections, for example, can be written as

$$\text{Rate} = n(Bn)(N_I - n_I) \qquad (6.11)$$

The Auger process is the inverse of impact ionization, in which a high-energy free carrier gives up its energy to a bound carrier to raise it to the band. By balancing impact ionization with Auger recombina-

Figure 6.3. Electron capture cross section of a particular defect in n-type GaAs as a function of temperature for seven samples with different electron densities.[4]

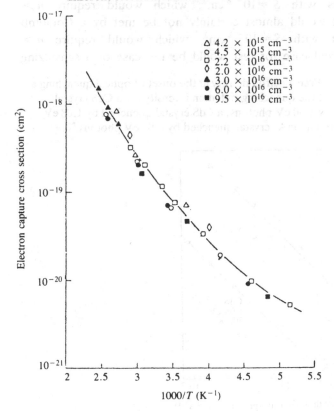

tion for a hydrogen-like center, a value for the coefficient $B$ can be calculated to obtain

$$B = (2.44 \times 10^{-19}/E_I T^2)(1/m_n{}^*)[(1 + 0.522 \log \epsilon)/\epsilon] \, \text{cm}^6 \, \text{s}^{-1} \qquad (6.12)$$

where $E_I$ is the ionization energy of the center at temperature $T$ in a semiconductor with effective electron mass $m_n{}^*$ and relative dielectric constant $\epsilon$.[5] Eq. (6.12) gives a value for $B \approx 10^{-24} \, \text{cm}^6 \, \text{s}^{-1}$.

The question arises as to how general this value of $B$ is, since it was calculated for the special case of a hydrogenic imperfection, but most imperfections important as recombination centers are not hydrogenic. A simple way to test this situation is to make use of the onset of optical quenching phenomena shown in Figures 5.10 and 5.11. Consider the condition for the competition between Auger recombination and normal recombination:

$$nSv_n(N_I - n_I) = n(Bn)(N_I - n_I) \qquad (6.13)$$

In order for Auger recombination to be dominating at the onset of optical quenching, $n > Sv_n/B$. This condition might be met for a radiative process with $S \approx 10^{-18} \, \text{cm}^2$, which would require $n > 10^{12} \, \text{cm}^{-3}$, but it would almost certainly not be met by a Coulomb attractive center with $S \approx 10^{-12} \, \text{cm}^2$, which would require $n > 10^{19} \, \text{cm}^{-3}$. An ideal case for test would be the case of a sensitizing

Figure 6.4. Free electron density at the onset of optical quenching as a function of the quenching radiation intensity for a CdS crystal quenched by 1.30 eV photons, a CdS crystal quenched by 1.08 eV photons, and a GaAs crystal quenched by 1.08 eV photons.[6]

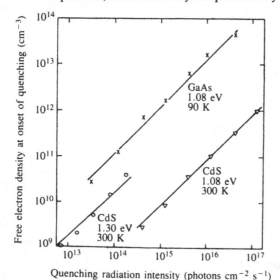

Table 6.1. *Summary of recombination at imperfections*

| Energy released as | Type of recombination center | Typical 300 K cross section, $S$, cm$^2$ |
| --- | --- | --- |
| Photons: $\Delta E = \hbar\omega_{pt}$ | Neutral or Coulomb repulsive | $10^{-19}$ |
| Phonons: $\Delta E = n\hbar\omega_{pn}$ (simultaneous) | Neutral | $\leq 10^{-16}$ |
| Phonons: $\Delta E = n\hbar\omega_{pn}$ (simultaneous) | Coulomb repulsive | $\leq 10^{-19}$ |
| Phonons: $\Delta E = n\hbar\omega_{pn}$ (sequential) | Coulomb attractive | $10^{-12}$ |
| Free carriers (Auger) | Neutral or Coulomb repulsive | $f(n)^a$ |

$^a$ Of the form $\beta_n = Bn$.

center with $S \approx 10^{-21}$ cm$^2$, which would require only that $n \geq 10^{10}$ cm$^{-3}$.

Optical quenching can be used to discriminate between Auger and normal recombination through a sensitizing center because of the different dependence of $n_b$ at the breakpoint and the intensity of the quenching light. The normal case is described by Eq. (5.11) and $n_b = F_Q S_0/\beta_n$. For the case of Auger recombination,

$$n_b B n_b (N_I - n_I) = F_Q S_0 (N_I - n_I) \tag{6.14}$$

and $n_b = (F_Q S_0/B)^{1/2}$. Therefore $n_b \propto F_Q$ should be linear at low quenching intensities, and then when $n$ increases beyond the critical value for Auger recombination to dominate, $n_b \propto F_Q^{1/2}$.

Measurements have been made for CdS crystals at 300 K and for GaAs crystals at 90 K. Figure 6.4 shows a plot of log $n_b$ vs log $F_Q$. No break from a $n_b \propto F_Q$ behavior is found for the GaAs crystal for values of $n_b$ as high as $4 \times 10^{13}$ cm$^{-3}$. Since for this case, $S_n = 10^{-21}$ cm$^2$, it follows that the value of $B < 10^{-28}$ cm$^6$ s$^{-1}$ for sensitizing centers in GaAs. This result indicates that Auger recombination is characterized by a lower value of $B$ for deep states than for shallow hydrogenic states.

## 6.5 Summary of recombination cross sections

The results of this chapter are summarized in Table 6.1.

# 7

# Steady state photoelectronic analysis

In this chapter and the next we consider a variety of methods for photoelectronic analysis of semiconductors. From such analysis we desire to obtain information about the density, ionization energy, electron capture cross section, hole capture cross section, and optical cross section of specific imperfections present in the material as grown or introduced deliberately by processing. In this chapter we focus on several steady state methods for obtaining this information; in Chapter 8 we look at several transient methods.

## 7.1 Temperature dependence of dark conductivity

The most straightforward electrical measurement capable of giving some information about imperfection properties is the measurement of the electrical conductivity as a function of temperature. For this purpose ohmic contacts are used if possible, so that the bulk conductivity of the material can be measured. If the contacts are in question, a four-point contact method can be used, in which current $I$ is passed through the two outer contacts and the voltage drop $\phi$ is measured on two inner contacts through which no current is drawn. If the distance between the inner contacts is $d$ and the cross sectional area is $A$, then

$$\sigma = (I/\phi)(d/A) \tag{7.1}$$

Since $\sigma = nq\mu$ for one-carrier conductivity, the measurement of the temperature dependence of $\sigma$ will depend both on the temperature dependence of $n$ and of $\mu$. If the temperature dependence of $\mu$ is known or can be measured separately (as, for example, by doing a Hall effect or thermoelectric power measurement – see below), then the temperature dependence of $n$ can be compared with simple expressions for the expectations for uncompensated or compensated donors or acceptors.

For example, if a material contains only donors with energy level $E_D$ above the valence band, then consideration of charge neutrality gives for the temperature dependence of the electron density at low temperatures in the range in which the donors are only partially ionized:

$$n = (N_D N_c/g)^{1/2} \exp[-(E_c - E_D)/2kT] \qquad (7.2)$$

where the degeneracy $g = 2$ for simple donors. A plot of $\ln(nT^{-3/4})$ vs $1/T$ therefore gives a straight line with slope proportional to one-half the ionization energy of the donor, $(E_c - E_D)/2$, and with intercept that depends on the donor density $N_D$.

As another example, consider a material that contains $N_D$ donors with ionization energy of $(E_c - E_D)$ partially compensated by $N_A$ acceptors $(N_D > N_A)$ with ionization energy of $E_A$. Then there are two ranges in a measurement of $n$ vs $T$ when the donors are only partially ionized. In the first low-temperature range, $n \ll N_A$, $(N_D - N_A)$, and

$$n = (N_c/g)[(N_D - N_A)/N_A] \exp[-(E_c - E_D)/kT] \qquad (7.3)$$

so that a plot of $\ln(nT^{-3/2})$ vs $1/T$ is proportional to $(E_c - E_D)$. In an intermediate temperature range, $N_A \ll n \ll (N_D - N_A)$, and

$$n = [(N_D - N_A)N_c/g]^{1/2} \exp[-(E_c - E_D)/2kT] \qquad (7.4)$$

so that a plot of $\ln(nT^{-3/4})$ vs $1/T$ is proportional to $(E_c - E_D)/2$. The intercepts in both cases depend on $(N_D - N_A)$. Examples for uncompensated acceptors or partially compensated acceptors can be described in the same way with appropriate change of symbols.

Although these analyses require the appropriately temperature-corrected electron density to be plotted vs $1/T$, approximate values for the ionization energy can be obtained even if simply $\ln \sigma$ is plotted vs $1/T$, neglecting the slower power-law temperature dependence of the mobility and the density of states compared to the exponential variation of carrier density with temperature.

Care must be taken, however, in simple interpretation of dark conductivity or carrier density plots vs $1/T$ on real materials for at least two reasons: (*a*) a complex set of different donor levels may give a measured $\ln \sigma$ vs $1/T$ that experimentally consists approximately of a single exponential over an appreciable temperature range and should not be interpreted as evidence of a single donor level, and (*b*) if donors and acceptors are close to being completely compensated (i.e.,

$N_D = N_A$) the Fermi level will be pinned halfway between the donor and acceptor level and the corresponding activation energy for a $\ln(nT^{-3/2})$ vs $1/T$ plot is a combination of donor and acceptor imperfection energies: $(E_D + E_A)/2$.

## 7.2    Dark Hall effect

One way to separate the carrier density and mobility in the expression for the electrical conductivity is to carry out a Hall effect measurement. A magnetic field $(B_z)$ is applied normal to the direction of an applied electric field $(\mathscr{E}_x)$, and the induced potential difference in the third orthogonal direction is measured as the Hall voltage. The applied magnetic field causes the moving carriers to depart from a straight-line path and to build up a transverse electric field $(\mathscr{E}_y)$ in the $y$-direction with no current flow in the $y$-direction.

If all of the carriers (e.g., electrons) are of the same type (the scattering relaxation time $\tau_{sc}$ does not depend on electron energy, and the equal-energy surfaces are spherical), then a simple expression can be obtained for the Hall coefficient:

$$R_H = \mathscr{E}_y/J_x B_z = \pm 1/nq \tag{7.5}$$

where $J_x$ is the electric current density in the $x$-direction. The positive sign is for the case when the free carriers are holes, and the negative sign for the case when the free carriers are electrons. Since a measurement of the Hall effect yields a value for the carrier density, combination with a measurement of conductivity yields the Hall mobility:

$$\mu_H = \sigma R_H = \pm \mu \tag{7.6}$$

If both electrons and holes are simultaneously present, the Hall field due to displaced electrons will tend to cancel the Hall field due to displaced holes. For normally used, relatively small magnetic field strengths,

$$R_H = (p\mu_p^2 - n\mu_n^2)/[q(p\mu_p + n\mu_n)^2] \tag{7.7}$$

$$\mu_H = \sigma R_H = (p\mu_p^2 - n\mu_n^2)/(p\mu_p + n\mu_n) \tag{7.8}$$

If exact values of $n$ and $\mu$ are required, then certain relatively small corrections to these expressions must be considered for normal semiconductors. In general Eqs. (7.5) and (7.6) become

$$R_H = \pm KM/nq \tag{7.9}$$

$$\mu_H = \pm KM\mu \tag{7.10}$$

The correction term $K$ is due to the fact that for most scattering processes $\tau_{sc}(E)$ is a function of electron energy and for averaging processes yield a correction factor. For low magnetic fields

$$K = \langle \tau_{sc}^2 \rangle / \langle \tau_{sc} \rangle^2 \tag{7.11}$$

For elastic scattering by acoustic lattice waves or by charged impurities in a semiconductor, values of $K$ are temperature independent. $K = 1.18$ for acoustic lattice scattering only, and 1.93 for charged impurity scattering only; if both types of scattering are present simultaneously values of $K$ lie between 1.05 and 1.93.[1] For inelastic scattering by optical lattice waves, the value of $K$ varies with temperature from about 1.07 to 1.23 between 80 and 400 K.[2]

The correction term $M$ differs from unity when the energy band structure of the semiconductor is not spherically symmetric. If the band structure is defined by equal energy surfaces defined by a transverse effective mass $m_t^*$ and a longitudinal effective mass $m_l^*$, so that $\gamma = m_l^*/m_t^*$, then

$$M = 3\gamma(\gamma + 2)/(2\gamma + 1)^2 \tag{7.12}$$

For Si, $M = 0.87$ and for Ge, $M = 0.795$.

With these correction terms, Eq. (7.8) becomes

$$\mu_H = \sigma R_H = (pK_p^2 M_p^2 \mu_p^2 - nK_n^2 M_n^2 \mu_n^2)/(pK_p M_p \mu_p + nK_n M_n \mu_n) \tag{7.13}$$

## 7.3 Dark thermoelectric effect

Another method used to separate the carrier density and mobility effects in making measurements of electrical conductivity as a function of temperature is the thermoelectric effect. If a small temperature gradient is imposed on a material, the carriers at the hot end diffuse toward the cold end faster than they are replaced. As a result a potential difference, called the thermoelectric power $\alpha$ is set up to maintain zero net current flow along the material. For a metal, $\alpha$ is of the order of magnitude of a few $\mu V \, K^{-1}$, whereas for a typical non-degenerate semiconductor $\alpha$ is of the order of $1 \, mV \, K^{-1}$.

For an n-type semiconductor

$$\alpha_n = -(k/q)[A + \ln(E_c - E_F)/kT] \tag{7.14}$$

whereas for a p-type semiconductor

$$\alpha_p = (k/q)[A + \ln(E_F/kT)] \tag{7.15}$$

The factor $A$ depends on the scattering process. In view of the fact that $\tau_{sc} \propto E^{-s}$ for acoustic lattice scattering ($s = \frac{1}{2}$) and for charged impurity scattering ($s = -\frac{3}{2}$), we can rewrite Eqs. (7.14) and (7.15):

$$\alpha_n = -(k/q)[(\tfrac{5}{2} - s) + \ln(N_c/n)] \tag{7.16}$$

$$\alpha_p = (k/q)[(\tfrac{5}{2} - s) + \ln(N_v/p)] \tag{7.17}$$

where $N_c$ and $N_v$ are the effective densities of states in the conduction and valence bands respectively. In the case of scattering by optical lattice waves, $A$ varies with temperature between 1.8 and 2.4 between 80 and 400 K, with its minimum value at about 170 K.[2]

Measurements of thermoelectric power therefore give a way to determine the carrier density; combined with measurements of electrical conductivity, both carrier density and mobility can be determined. If measurements of conductivity, Hall effect, and thermoelectric effect are carried out on the same material, comparison provides information about the type of scattering.

## 7.4    Dark Schottky barrier capacitance

A third major technique for separating carrier density and mobility effects is the measurement of the capacitance $C$ of a Schottky barrier on the semiconductor as a function of applied bias voltage $\phi_a$. The relationship for this measurement has already been given in Eq. (1.26) for an n-type semiconductor with $N_D^+$ ionized donors:

$$(C/A)^{-2} = [2(\phi_D - \phi_a)/(\epsilon_r \epsilon_0 q N_D^+)] \tag{1.26}$$

A plot of $(C/A)^{-2}$ vs $\phi_a$ gives a straight line with a slope that varies inversely with ionized donor density, assumed in this case to be equal to the free electron density in the semiconductor bulk. Other quantities that need to be known are $A$, the area of the junction, and $\epsilon_r$, the relative dielectric constant of the semiconductor.

In a more general case the density entering Eq. (1.26) is the net positive charge in a region of material in which both positively charged and negatively charged imperfections may be present; even in this case it can be considered to be effectively equal to the free carrier density (associated with uncompensated imperfections) in the bulk.

Measurements of electrical conductivity, Hall effect, and thermoelectric power are all much more accurate when ohmic contacts are made to the semiconductor. If this is difficult, the use of the capacitance measurement allows determination of the free carrier density using a non-ohmic contact as a Schottky barrier.

## 7.5    Optical absorption

The measurement of the optical absorption constant $\alpha$ as a function of photon energy is a standard method for the determination of imperfection levels in a semiconductor. The absorption shows an edge corresponding to the minimum absorption transition from the imperfection to one of the bands. Practical problems occur for very high absorption constants and bulk samples, and in such cases measurements of reflectivity are often used instead. Practical problems also occur for very low absorption constants in thin film materials.

An ingenious method for the measurement of absorption corresponding to low absorption constants in thin films is the method known as photodeflection spectroscopy.[3] When an ac (chopped) light is absorbed by a material, periodic heating of the material is caused. If this material is immersed in a suitable liquid such as $CCl_4$, then the temperature of the liquid and hence its index of refraction is changed. A laser beam passing the material at a grazing angle is deflected by the change in optical properties of the liquid, and the magnitude of the deflection of the laser beam can be directly correlated with the absorption constant of the film. Absorption constants in the range of $1-10^2\,cm^{-1}$ can be readily measured with films of the order of $1\,\mu m$ in thickness.

## 7.6    Photoconductivity

Steady state photoconductivity depends directly on the majority carrier lifetime, as for example in n-type material:

$$\Delta n = G\tau_n \qquad (7.18)$$

Measurement of the photoexcitation rate, from a knowledge of absorption constant and incident photon flux (see Section 2.3) and of the photoconductivity $\Delta\sigma$ allows the calculation of the electron lifetime, provided that reasonable assumptions can be made about the magnitude of the electron mobility and the effects of photoexcitation, if any, on it.

Measurement of the photoconductivity as a function of photoexcitation intensity to determine the value of $\gamma$ in $\Delta n \propto G^\gamma$ provides background information that can be used in testing a model of the process, as it involves one or more traps and one or more recombination centers.

If the semiconductor contains sensitizing-type imperfections, as discussed in Chapter 5, then measurements of photoconductivity as a function of intensity and temperature can yield the data of supralinear photoconductivity and thermal quenching of photoconductivity that can be analyzed to obtain information about the location of the imperfection energy level and the ratio of its capture cross sections. If saturation of photoconductivity can be observed, the density of the sensitizing centers can be deduced. Optical quenching of photoconductivity is able to provide additional information about the location of the imperfection energy level, and the value of its majority carrier capture cross section.

Photoconductivity measurements under conditions in which the increase in carrier density $\Delta n$ is much larger than the dark electron density $n_0$ can readily be carried out with constant illumination. In cases, however, where $\Delta n < n_0$, it is usually necessary to use an alternating (chopped) illumination so that ac amplification techniques with a lock-in amplifier can be used to make the small $\Delta n$ measurable.

The variation of photoconductivity with incident photon energy, yielding an excitation spectrum for photoconductivity, can be used to detect the presence of imperfection levels and to determine their location. Photoconductivity measurements can also be used to determine the density of imperfections in a spectral region where the optical absorption constant due to the relatively small imperfection density is too small for standard optical transmission measurements. A particularly striking example is the development of the so-called constant-photocurrent measurement (CPM) for the determination of the defect density (of the order of $10^{16}$ cm$^{-3}$ – corresponding to an absorption constant of about $1$ cm$^{-1}$) in amorphous Si films with thickness of about $1 \mu m$.[4] In order to avoid changes in photoconductivity with changes in photoexcitation rate as the spectrum is scanned, the measurements are made with a constant photocurrent (and hence a constant steady state Fermi level) achieved by changing the intensity of the light, which becomes the experimental variable.

## 7.7    Minority carrier lifetime

The measurement of the minority carrier lifetime must, of necessity, be somewhat more indirect than the measurement of the majority carrier lifetime. A variety of different methods has been developed with specific applicability for different materials and conditions.[5]

### Photoconductivity saturation with electric field

Consider a situation like that of Figure 2.5($e$). The contacts are such that they are either blocking, or if not, measurements are for voltages below the range giving rise to space-charge-limited currents. Absorption creates an electron and a hole; it is assumed that electrons are the majority carriers, and that holes are the minority carriers. When the applied voltage is high enough to sweep the hole out of the sample before it can be captured at an imperfection ($\phi = \phi_c$), the photoconductivity saturates. If the contacts are ohmic, the gain changes from ($\tau_n \mu_n \phi / d^2$) for $\phi < \phi_c$ to ($1 + \mu_n / \mu p$) for $\phi > \phi_c$. If the contacts are blocking, saturation corresponds to achieving a gain of approximately unity, limited at lower $\phi$ by trapping. The transit time for the hole is

$$t_{tr} = d^2 / \mu_p \phi_c = 1/\beta_p n_I = \tau_p \qquad (7.19)$$

where the width of the sample is approximately $2d$, $\phi_c$ is the critical applied voltage for the onset of photoconductivity saturation, $\beta_p$ is the hole capture coefficient of the imperfections, and $n_I$ is the density of electron-occupied imperfections. It follows from Eq. (7.19) that

$$\tau_p \mu_p = d^2 / \phi_c = d / \mathscr{E}_c \qquad (7.20)$$

where $\mathscr{E}_c$ is the critical electric field for saturation.

If the imperfection is a sensitizing center with a Coulomb attractive cross section for holes, then $d$ must be fairly small to achieve photoconductivity saturation (of the order of $1\,\mu$m). But if the imperfection has a neutral cross section for holes, then the value of $d$ can be as large as 1 cm and still permit observation of saturation.

### Photomagnetoelectric effect

The photomagnetoelectric effect is the Hall effect of carriers diffusing away from a surface where they have been created by strongly absorbed light, as pictured in Figure 7.1.[6] It is closely related to the Dember effect described in Section 3.4. For small values of magnetic field, $B$, the horizontal deflection of electrons, for example, is

$$\delta_n = r\theta \approx (D_n \tau)^{1/2} (\mu_n B) \qquad (7.21)$$

where the diffusion length of electrons is $L_n = (D_n \tau)^{1/2}$, $\tan \theta = \mu_n B = \theta$, and $\tau$ is the minority carrier lifetime. The charge contribution to the external circuit is given by

$$J_{PME} = q(\delta_n / d + \delta_p / d) F \qquad (7.22)$$

where $F$ is the incident photon flux per unit area per unit time.

Substituting Eq. (7.21) into (7.22), with $D = \mu kT/q$, gives

$$J_{PME} = (FBkT/q)(\mu_n^{3/2} + \mu_p^{3/2})\tau^{1/2} \qquad (7.23)$$

In the same sample the photocurrent density can be written

$$J_{PC} = [qF\tau_{maj}(\mu_n + \mu_p)\phi]/d^2 \qquad (7.24)$$

In the fortuitous case that $\tau_{maj} = \tau_{min} = \tau$,

$$J_{PC}/J_{PME} = (\phi/dB)(q/kT)^{1/2}(\tau/\mu^*)^{1/2} \qquad (7.25)$$

where

$$1/\mu^{*1/2} = (\mu_n + \mu_p)/(\mu_n^{3/2} + \mu_p^{3/2}) \qquad (7.26)$$

The variation of $J_{PME}$ with $B$ is strongly dependent on the surface recombination velocity: if the surface recombination velocity is high, $J_{PME}$ shows a maximum for a particular value of $B$, whereas if the surface recombination velocity is low, $J_{PME}$ saturates at a maximum value with increasing $B$.

### Surface photovoltage

The absorption of light at a semiconductor surface reduces the band bending at that surface, and the photoinduced voltage indicates

Figure 7.1. Schematic diagram for the photomagnetoelectric effect with strongly absorbed light incident in the $y$-direction.

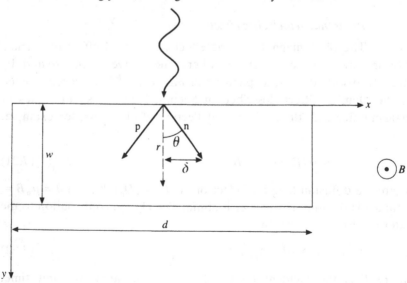

the direction and magnitude of the band bending at the surface. In order to measure the minority carrier diffusion length (and hence the minority carrier lifetime) the surface of the material is illuminated with monochromatic light with different photon energies and hence different values of absorption constant, near the absorption edge. The range of photon energies is chosen so that the penetration depth of the light is much larger than the space charge region near the surface, but much less than the thickness of the material. A photovoltage is developed between the surface and the bulk due to the back diffusion of minority carriers to the surface barrier and the subsequent charge separation of the hole–electron pairs in the barrier field.[7,8]

If experimentally the surface photovoltage is kept constant by varying the photon flux $F(\lambda)$,

$$F(\lambda) = \text{const}[1/\alpha(\lambda) + L_{min}] \tag{7.27}$$

then a plot of $F(\lambda)$ vs $1/\alpha(\lambda)$ produces a straight line with intercept on the negative $1/\alpha(\lambda)$ axis of $L_{min} = (\mu_{min}\tau_{min}kT/q)^{1/2}$.

### Electron- and photon-beam-induced current

In this method for determining the minority carrier lifetime, a small region of the material is scanned with an electron beam (EBIC) or a light beam as a function of distance $x$ from a p–n junction or a Schottky barrier involving the material. The collected current $J_{sc}$ is a complex function of geometry and materials' properties, but for moderate to large values of $x/L_{min}$, it can be expressed fairly simply as

$$J_{sc} = A \exp(-x/L_{min}) \tag{7.28}$$

where $A$ is a constant. A plot of $\ln J_{sc}$ vs $x$ yields a straight line with slope proportional to $L_{min}$.[9] The results can be strongly affected by surface recombination effects.

### Spectral response of junctions

Collection of electron–hole pairs by a semiconductor junction depends on the distance that these carriers are photocreated from the junction itself. If the distance is much larger than a diffusion length, recombination occurs before the carriers are collected. If a semiconductor junction can be prepared involving the material under investigation as the main absorber of light, then the dependence of the short-circuit current $J_{sc}$ on photon energy (with the accompanying variation in absorption constant) can be modelled in such a way that the value of $L_{min}$ can be deduced from the best fit to the data.[10]

### Steady-state photocarrier grating technique

A relatively recently developed technique makes it possible to measure the ambipolar diffusion length under steady state conditions in insulating photoconductive materials.[24] Interference between two coherent light beams with different intensities is achieved, in order to create a grating superposed on a background uniform illumination. If the grating spacing is much larger than the diffusion length, a well-defined photocarrier grating is created, but if the grating spacing is much less than the diffusion length, an effectively uniform carrier distribution results. Measurement of the photocurrent perpendicular to the grating fringes as a function of the grating period permits determination of the diffusion length. Diffusion lengths between 20 and $10^4$ nm can be accurately measured.

### Transient Methods

Two additional transient methods for determining the minority carrier lifetime have been frequently used. Since they are not steady state methods, they do not strictly belong in this chapter. They involve the time constant for the decay of an open-circuit voltage after stopping photoexcitation,[11] and the time constant of the capacitance decay of a metal–insulator–semiconductor (MIS) junction after the application of a large reverse-bias step,[12] which is the most sensitive method, allowing measurement of $\tau_{min}$ as low as $10^{-9}$ s, providing that the structure required can be prepared with the semiconductor under investigation.

## 7.8    Photo-Hall effects

As the Hall effect can be used to separate carrier density and mobility effects under thermal equilibrium conditions, the photo-Hall effect can be used to separate the effects of photoexcitation on carrier density and mobility.[13-17] Mobility changes under photoexcitation occur either because of a change in the charge on charged impurities in a temperature range where charged impurity scattering is important in determining the mobility, or because of inherent inhomogeneities in the material that are reduced by photoexcitation. In addition, the photo-Hall effect is extremely useful under situations in which the dominant carrier actually changes under photoexcitation, as for example when a material that is p-type in the dark becomes n-type in the light. A few examples are given in the following to indicate the kind of results that can be obtained.

*Change in scattering cross section by photoexcitation*

If a material has scattering by charged impurities, as well as from other mechanisms, we may write in the dark

$$1/\mu_d = 1/\mu_{CI} + 1/\mu_{oth} \tag{7.29}$$

where $\mu_{CI}$ is the mobility due to charged impurity scattering, and $1/\mu_{oth}$ is the mobility due to all other scattering processes. If, for simplicity, we consider the case where photoexcitation removes all of the charged impurity scattering but does not affect other scattering processes (and if we neglect scattering due to neutral impurities), then in the light $1/\mu_l = 1/\mu_{oth}$ and we obtain by subtraction that

$$\Delta(1/\mu) = (1/\mu_d - 1/\mu_l) = 1/\mu_{CI} = m^*NSv/q \tag{7.30}$$

where $N$ is the density of charged impurities, $S$ is the scattering cross section of the charged impurities, $m^*$ is the effective mass of the corresponding carriers, and $v$ is the thermal velocity of the carriers. Clearly this expression can be modified if only a fraction of the charged impurities is neutralized, or if the change in cross section

Figure 7.2. Effects of photoexcitation on the mobility in Si:Zn crystals.[18]

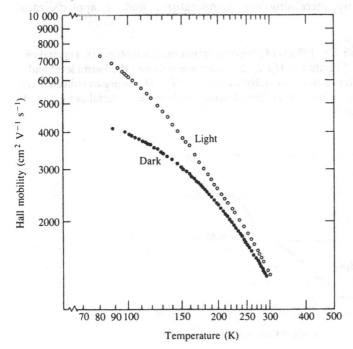

Hall mobility (cm$^2$ V$^{-1}$ s$^{-1}$)

Temperature (K)

corresponds to a change from doubly charged to singly charged, for example. Figure 7.2 shows data for the temperature dependence of the electron mobility in Zn-doped Si in the dark and light;[18] the results are consistent with the known density of Zn and a change in charge from −2 to −1 under photoexcitation.

When photo-Hall measurements are made on a variety of other II–VI and III–V semiconductors in a high-resistivity state, however, a much larger increase in mobility is found with photoexcitation than is predicted by these considerations using a standard Coulomb scattering cross section. Figure 7.3 is a plot of $1/\mu$ vs the electron Fermi level as the latter is varied by photoexcitation for a crystal of CdS. The data are for a case where the electron mobility at room temperature is observed to increase from a value of $110 \text{ cm}^2 \text{ V}^{-1} \text{ s}^{-1}$ corresponding to $n = 5 \times 10^8 \text{ cm}^{-3}$ in the dark, to $300 \text{ cm}^2 \text{ V}^{-1} \text{ s}^{-1}$ corresponding to $n = 2 \times 10^{14} \text{ cm}^{-3}$ in the light. Three distinct steps are indicated in Figure 7.3. Independent knowledge of the density and depth of three imperfection levels in the material allows a good reproduction of these results under the assumption that their charge is removed when the Fermi level rises through them, and the assumption that their effective scattering cross section is 100 times larger than a Coulomb cross section. The magnitude of these steps in mobility with photoexcitation is reduced by increasing the temperature, and it appears most

Figure 7.3. Effects of photoexcitation on the mobility in a crystal of CdS.[14] Plotted as $1/\mu$ vs the difference between the Fermi level and the conduction band edge with $\beta = m^*/q$. The energies indicated are the energies of three imperfection levels in the material as measured independently.

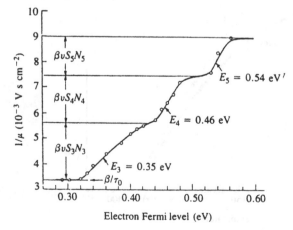

reasonable to assign their existence to scattering by space-charge regions in the crystal associated with inhomogeneous imperfection distributions.

### Effects associated with optical quenching

Optical quenching of photoconductivity removes holes from sensitizing centers and thereby increases their charge. Photo-Hall measurements during optical quenching indicate that both in-homogeneity effects and initiation of two-carrier effects may be important depending on the situation.

Figure 7.4. Results of photo-Hall measurements during optical quenching of photoconductivity in a CdS crystal.[14] (a) Per cent of the original photocurrent as a function of the quenching intensity. (b) Measured Hall mobility under optical quenching $\mu_{Q+L}$ as a function of the per cent of the original photocurrent. (c) Measured Hall mobility with decreasing primary light intensity in the absence of optical quenching $\mu_L$, compared with the measured Hall mobility with constant primary photoexcitation and increasing quenching intensity as a function of electron density. (d) The ratio of the two mobilities plotted in (c).

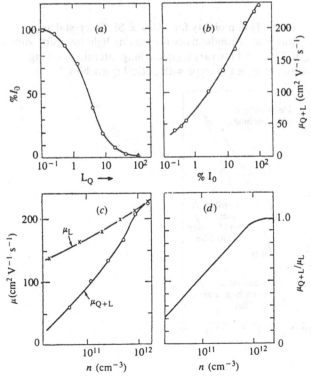

Figure 7.4 shows the effects of optical quenching on the Hall mobility in a CdS crystal. The mobility decreases from $235\,\mathrm{cm^2\,V^{-1}\,s^{-1}}$ in the absence of optical quenching to $41\,\mathrm{cm^2\,V^{-1}\,s^{-1}}$ for 99.8% quenching. The data shown in Figure 7.4 sort out the effects on the mobility due simply to a change in the Fermi level, as described in the previous section, and indicate that the effects of optical quenching are much larger. Either inhomogeneity effects dominate, giving an effective scattering cross section several orders of magnitude larger than a Coulomb cross section, or optical quenching has induced two-carrier effects such that the density of free holes in the strongly quenched range (greater than a factor of 5 in mobility) must be at least an order of magnitude larger than the density of free electrons.

In the results of photo-Hall measurements on a GaAs:Si:Cu crystal shown in Figure 7.5, it is clear that optical quenching has changed the conductivity type from n-type to p-type. Again inhomogeneous scattering effects due to the change in Fermi level alone are indicated, in addition to the initiation of two-carrier conductivity.

Figure 7.5. Photo-Hall mobility for a GaAs:Si:Cu crystal as a function of illuminated conductivity for varying light intensity without quenching at 77 K, and for varying quenching intensity showing conversion from n-type to p-type with optical quenching.[14]

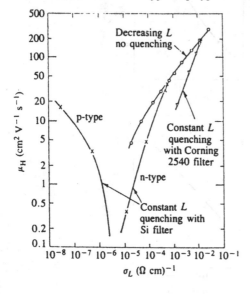

### Effects associated with temperature dependence of lifetime

In some cases conversion of conductivity type similar to that just described for optical quenching can occur as the result of temperature effects. Figure 7.6 shows the temperature dependence of the dark conductivity, the photoconductivity, the dark Hall mobility and the photo-Hall mobility for a crystal of GaAs:Si:Cu. In the dark the conductivity is p-type. The crystal shows high photosensitivity at low temperatures where the photoconductivity is n-type, but undergoes thermal quenching at about −60 °C at the light level used and the photoconductivity converts to p-type at higher temperatures. These results are the consequence of sensitized photoconductivity associated with sensitizing centers lying 0.45 eV above the valence band; when holes can be stably held in these centers, the electron lifetime increases by a large factor. In the temperature range between −50 and +20 °C, it was possible to measure the small Hall mobility only sporadically; this effect is probably associated with an inhomogeneous conversion of the material from p-type to n-type.

Rather different variations of photoconductivity type with temperature are shown in Figure 7.7 for a Bridgman-grown crystal of GaAs

Figure 7.6. Temperature dependence of photoconductivity, dark conductivity, dark Hall mobility and photo-Hall mobility for a GaAs:Si:Cu crystal.[14]

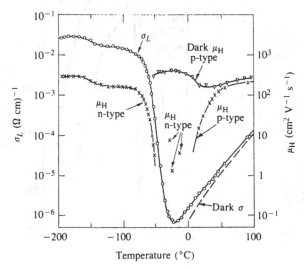

annealed for 16 hours at 450 °C in the presence of Cu. Figure 7.7(*a*) shows the variation of photo-Hall mobility with light conductivity, varied by increasing photoexcitation intensity, for each of three different temperatures. For low intensities the photoconductivity is n-type, then converts to p-type for intermediate intensities, and finally converts back to n-type for high intensities. The effect can be associated with the presence of a high density of traps with depth of about 0.56 eV that trap electrons effectively in the intermediate intensity range.

Figure 7.7(*b*) shows the temperature dependence of photoconductivity and photo-Hall mobility for the same crystal for the maximum intensity used in Figure 7.7(*a*). Upon cooling, the n-type Hall mobility rises steeply, reaches a maximum value, and then drops off again as at low temperatures the photoconductivity abruptly changes to p-type. The rapid increase in n-type Hall mobility upon cooling, in view of the existence of a two-carrier regime, suggests the existence of a hole trap

Figure 7.7. Variations in photo-Hall mobility for a crystal of GaAs annealed in the presence of Cu. (*a*) As a function of light conductivity varied by changing the light intensity at different temperatures. (*b*) Together with the light conductivity as a function of temperature.[14]

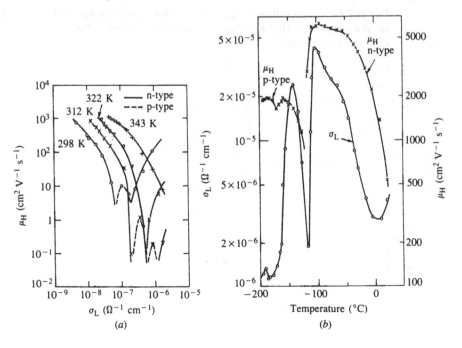

with depth between 0.4 and 0.5 eV which reduces the free-hole density, as the 0.56 eV electron trap had reduced the free electron density at room temperature. The sudden change from n- to p-type conductivity suggests the presence of a shallow electron trap that removes free electrons from the conduction band at low temperatures. Clearly all insights into the n–p conductivity transitions and their variation with intensity and temperature depend on the ability to make photo-Hall measurements.

Figure 7.8. (*a*) Photoconductivity spectral response measured at 82 K for high-resistivity GaAs measured from high to low photon energies, reading after 3 min (open circles) and after 15 min (open triangles); measured from low to high photon energies, reading after 3 min (filled circles) and after 15 min (filled triangles). (*b*) Photo-Hall mobility vs photon energy for the same symbols as in (*a*).[17]

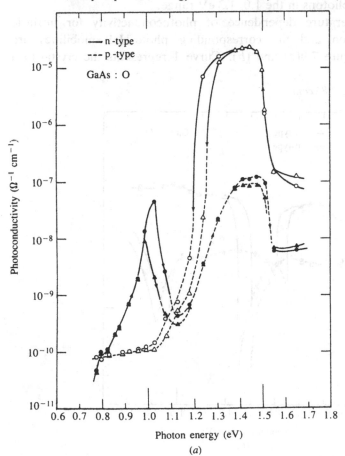

(*a*)

### Persistent effects in high-resistivity GaAs

A striking example of the value of photo-Hall measurements is given in Figures 7.8 and 7.9 for data taken with a crystal of high-resistivity GaAs grown with incorporated oxygen. Figure 7.8(*a*) and (*b*) shows the spectral response for photoconductivity and photo-Hall mobility as measured 3 and 15 min after onset of photoexcitation at each selected photon energy in two different ways; (i) measuring from high to low photon energy on a previously heated and cooled (in the dark) crystal, and (ii) measuring from low to high photon energy on a previously heated and cooled (in the dark) crystal. The results indicate two 'states' for the crystal: a higher-sensitivity n-type state, encountered if the sample has not been previously exposed to photons in the 1.0–1.3 eV range, and a lower-sensitivity p-type state to which the higher-sensitivity n-type state is reduced by exposure to photons in the 1.0–1.3 eV range.

The temperature dependence of photoconductivity for intrinsic photoexcitation and the corresponding photo-Hall mobilities are shown in Figure 7.9(*a*) and (*b*). Curve 1 represents the crystal in a

Figure 7.8 (*cont.*)

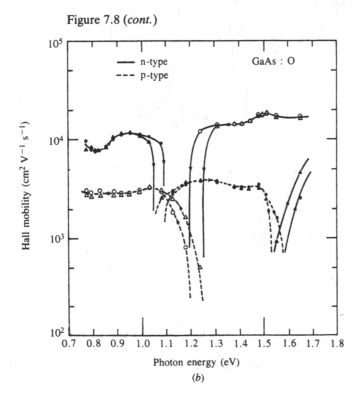

(*b*)

high-sensitivity state, and curves 2 and 3 represent the crystal in a low-sensitivity state at 82 K. This quenching process is persistent at low temperatures, and the low-sensitivity quenched state remains unchanged long after the removal of the quenching radiation even with intrinsic photoexcitation present. Upon heating, however, a dramatic recovery from the low-sensitivity quenched state to the high-sensitivity state occurs when the temperature exceeds 105 K. Curve 4 shows the

Figure 7.9. (*a*) Temperature dependence of photoconductivity for crystal of Figure 7.8 after quenching 1 h with 1.08 eV (circles) or 1.18 eV (squares) photons at 82 K, measured while warming; and for simultaneous intrinsic and 1.18 eV photoexcitation (inverted triangles). (*b*) Temperature dependence of the photo-Hall mobility for the same symbols as in (*a*).[17]

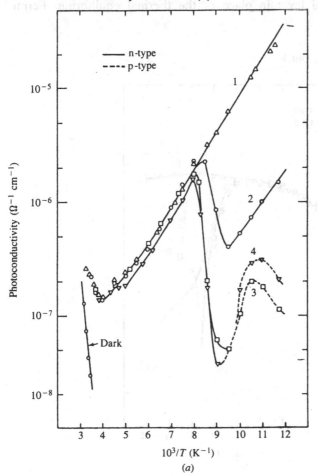

(*a*)

temperature variation for simultaneous intrinsic photoexcitation and
1.18 eV irradiation during heating.

## 7.9    Photothermoelectric effects

The photothermoelectric effect plays the same role for separating carrier density and mobility effects, and determining the carrier
type, under photoexcitation as does the photo-Hall effect. In addition
it has the advantage of requiring simpler apparatus, being suitable for
use with smaller mobility materials, and being more sensitive to
two-carrier effects. The expression for the thermoelectric power in the
dark involves the ratio $(E_c - E_F)/kT$ for an n-type material; the
expression for the thermoelectric power under photoexcitation is still
valid provided that the ratio is replaced by $(E_c - E_{Fn})/kT$, involving the
steady state Fermi level in place of the thermal equilibrium Fermi
level.

Figure 7.9 (*cont.*)

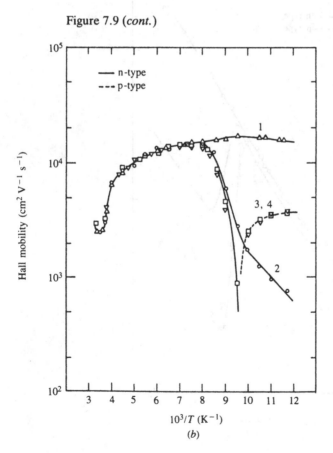

(*b*)

Figure 7.10 compares photo-Hall mobility values and photothermoelectric power values for the same sample measured over the same range of light conductivities, and Figure 7.11 compares values of electron density determined from photo-Hall and photothermoelectric measurements independently. Because of the way the electron density enters the thermoelectric power in a logarithmic term, comparison of electron densities in this way is a very sensitive test of the applicability of the interpretive analysis. Similar good agreement was found for measurements comparing intrinsic photoexcitation (530 nm) and extrinsic photoexcitation (650 nm). Like the photo-Hall effect, the photothermoelectric effect can be used to follow a variety of other effects such as optical quenching of photoconductivity (see Figure

Figure 7.10. A comparison of the photo-Hall mobility (*a*) and the photothermoelectric power (*b*) as a function of light conductivity for three different temperatures for a CdS crystal.[19]

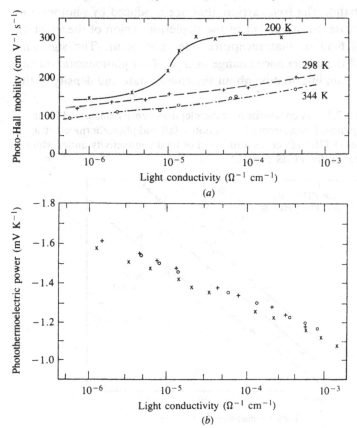

7.12), and even photoconductivity decay with time after stopping photoexcitation (see Figure 7.13). The photothermoelectric effect can also be used in two-carrier situations, where its value is given by

$$\alpha = (-k/q)\{n\mu_n[(5/2-s) + \ln(N_c/n)]$$
$$- p\mu_p[(5/2-s) + \ln(N_v/p)]\}/(n\mu_n + p\mu_p) \quad (7.31)$$

## 7.10   Photocapacitance

According to Section 7.4, the dark capacitance of a Schottky barrier on a material depends upon the uncompensated charge in the depletion layer. This uncompensated charge can be altered by photoexcitation, either decreasing the positive charge by exciting from the valence band to empty donor levels, increasing the positive charge by exciting from deep donors to the conduction band, increasing the negative charge by exciting from the valence band to empty acceptors, or decreasing the negative charge by exciting from acceptors to the conduction band. The free carriers that are produced by photoexcitation in each case are swept out of the depletion region of the junction by the local field so that recapture does not occur. The sign and magnitude of the capacitance change as a result of photoexcitation can be used to obtain information about the charge state and density of the

Figure 7.11. A comparison of the electron density calculated from independent measurements of photo-Hall and photothermoelectric power (PTEP) effects as a function of light conductivity under white excitation for a CdS crystal.[19]

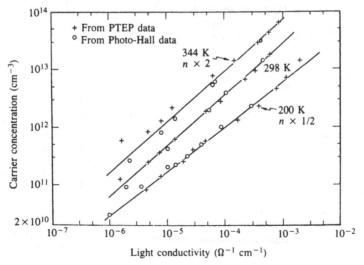

imperfection levels in the depletion layer. Since the initial rate of change of charge in the jucntion $Q$ is given by

$$dQ/dt = S_oN_IF \qquad (7.32)$$

where $S_o$ is the optical cross section of a particular set of imperfection levels with density $N_I$, and $F$ is the incident photon flux per unit area per unit time, measurements of the time dependence of the capacitance can be used to determine values for $S_o$ as a function of photon energy if the density $N_I$ is known.

Typical results are shown in Figure 7.14 for a Schottky barrier made by depositing Au on a liquid-phase epitaxial (LPE) layer of GaAs grown with Cu in the melt.[20] The diode was first forward biased at 77 K

Figure 7.12. Use of photothermoelectric effect to measure optical quenching in CdS. (a) Optical quenching of photoconductivity vs photon energy of quenching light. (b) Comparison of measured thermoelectric power (data points) with thermoelectric power calculated from the photoconductivity data of (a) (dashed line).[19]

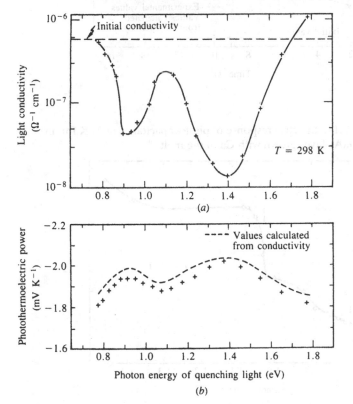

Figure 7.13. Use of photothermoelectric effect to measure decay of photoconductivity in CdS. (*a*) Current vs time during the decay of photoconductivity after reducing the photoexcitation intensity by a factor of $10^3$ for a photosensitive CdS crystal. (*b*) Comparison of the measured thermoelectric power (dashed line) with thermoelectric power calculated from the photoconductivity data of (*a*) (plus sign points).[19]

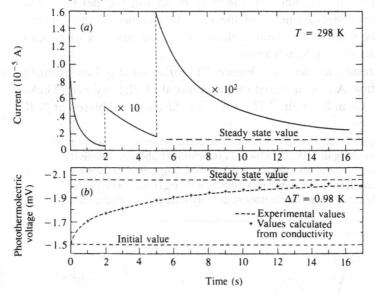

Figure 7.14. Spectral response of photocapacitance at 77 K for an LPE GaAs layer grown with Cu in the melt.[20]

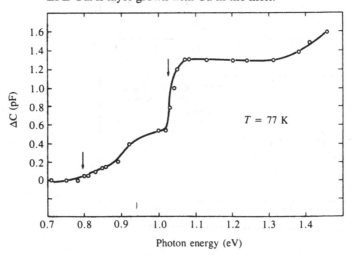

in the dark to fill all the centers. Then the diode was reverse biased and illuminated with light as a function of photon energy while the capacitance at 1 MHz was measured. Three characteristic thresholds are indicated in Figure 7.14, at about 0.80, 1.05 and 1.35 eV, which were identified respectively with a commonly found defect in GaAs lying 0.80 eV below the conduction band, a level associated with Cu lying 0.45 eV above the valence band, and the onset of intrinsic excitation. In order to determine the energy depth of the 0.80 and *1.05* eV transitions more exactly, the optical cross section was determined from photocapacitance transients according to Eq. (7.32), with the results given in Figure 7.15, where they are compared with the theoretical model due to Lucovsky.[21]

Two other photocapacitance-related phenomena were observed with these same layers. The first of these is not properly a steady state measurement, but we include it here because of its connection with the previous results. The diode was illuminated by intrinsic light with a

Figure 7.15. Spectral dependence of the optical cross section for two levels indicated in Figure 7.14.[20] Solid curves are theoretical curves.[21]

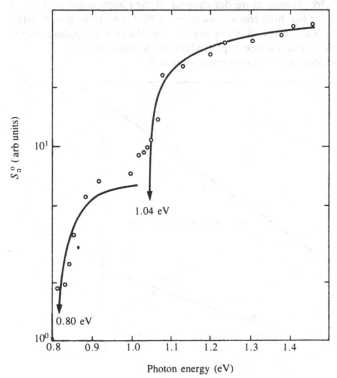

Photon energy (eV)

reverse bias applied in order to inject minority carriers (holes) into the depletion layer, where they are captured by hole traps. The capacitance transient, associated with thermal emptying of these traps after cessation of photoexcitation, was measured as a function of temperature. Figure 7.16 gives the variation of $\ln \tau_c$, where $\tau_c$ is the time constant of the capacitance transient, with $1/T$. Two hole traps are indicated, lying 0.20 eV and 0.43 eV (the Cu level) above the valence band, and with hole capture cross sections of about $10^{-21}$ and $10^{-16}$ cm$^2$ respectively.

It was also possible to measure the optical quenching of the thermally stable photocapacitance in these junctions at low temperature. Figure 7.17 shows that strong quenching has a threshold at 0.65 eV, indicating a deep hole trap at this energy above the valence band. At higher energies, excitation from levels lying 0.7 eV below the conduction band to the conduction band begins to dominate; these transitions cause an increase in photocapacitance and counteract quenching of the persistent capacitance.

Figure 7.16. Temperature dependence of the photocapacitance decay constant for two hole traps in an n-type LPE GaAs layer grown with Cu in the melt. The activation energy gives the hole ionization energy of the level, and the intercept enables one to calculate the corresponding hole capture cross section.[20]

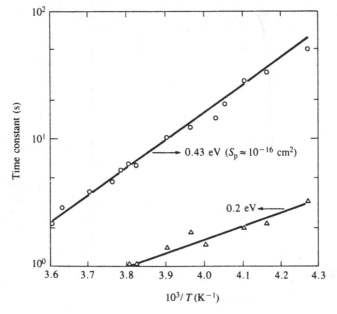

## 7.11  Luminescence

Measurements of the excitation and emission spectra for luminescence are also able to contribute strongly to the photo-electronic analysis of semiconductors. Since no electrical contacts are required, luminescence has been one of the most widely used methods for analysis over many years. On the other hand, since luminescence gives only the energy difference between a high-energy and a low-energy level, it does not specify by itself where these levels lie in the band diagram, or whether free carriers are involved or not. Most commonly luminescence is excited using photons, usually today from a laser, but it can also be excited by electrons, X-rays, or nuclear radiation, or in the presence of a semiconductor junction structure by forward bias of carriers followed by their radiative recombination.

A classical example of luminescence illustrating a few of the possibilities is given in the effects of Cu, Ag, and Mn impurities in ZnS. Emission spectra are given in Figure 7.18. The emission with maximum at 460 nm, labeled simply 'ZnS', results from heat treatment of ZnS with a halogen such as Cl; the luminescence is due to radiative recombination at a complex consisting of a halogen impurity on a S site and a neighboring Zn vacancy. The emissions with maximum at 455 nm due to Ag and at 530 nm due to Cu are caused by radiative recombination of a free electron (or possibly a localized electron in a

Figure 7.17. Optical quenching spectrum for photocapacitance at 77 K in an LPE GaAs layer grown with Cu in the melt. The initial persistent capacitance before optical quenching is 0.6 pF.[20]

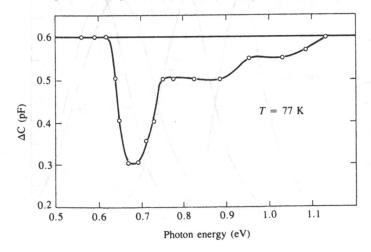

shallow level near the conduction band edge) with a hole on the Ag impurity level or the Cu impurity level. The emission with maximum at 590 nm is associated with Mn impurity and results from a transition from the first excited state of the $Mn^{2+}$ ion to the ground state of the ion.

Luminescence can be characterized by many of the phenomena previously discussed for conductivity or photoconductivity: an excitation spectrum can be measured, and thermal and optical quenching may occur. Specific emission bands may be identified with specific defects or impurities, thus making it possible to use luminescence as a diagnostic tool. Low-temperature luminescence may be characterized by details dependent upon excitonic recombination, which, in turn, may be dependent upon specific impurities, crystalline quality etc.

The excitation spectra for the emission spectra shown in Figure 7.18 are given in Figure 7.19. Each curve in Figure 7.19 is measured by monitoring the intensity at the emission maximum as a function of the exciting photon energy. In Figure 7.19 direct excitation from Ag and Cu levels lying in the bandgap is seen (bandgap of ZnS is about 3.7 eV) as well as direct excitation within the Mn ion to four excited states. If photoconductivity excitation spectra are measured on these materials, it is found that the excitation spectra for luminescence for Ag and Cu correspond to the excitation of n-type photoconduc-

Figure 7.18. Luminescence emission spectra for ZnS, ZnS:Ag, ZnS:Cu and ZnS:Mn samples.

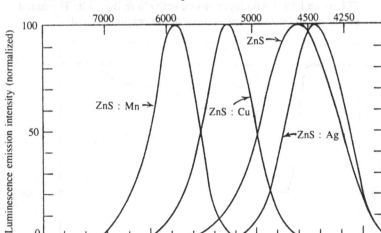

tivity, but that the excitation spectrum for the excited state transitions of Mn does not.

### Luminescence analysis of ZnSe:Cu crystals

Some interesting illustrations of the use of luminescence are provided by research on ZnSe:Cu crystals.[22] This is a system characterized by p-type dark conductivity, but n-type photoconductivity at lower temperatures due to the Cu behaving as a sensitizing center. At 77 K there are two luminescence bands, as shown in Figure 7.20: a red band with maximum at 1.97 eV and a green band with maximum at 2.34 eV. If the luminescence emission is measured at 16 K, another red band is found with maximum at 1.95 eV, which by its temperature behavior was identified with a pair transition involving 0.012 eV shallow levels.[23] At 77 K electrons captured in these shallow levels are no longer stably held long enough for the recombination transition to occur, and so it is replaced by a free electron to localized hole transition, the 1.97 eV emission.

Figure 7.19. Excitation spectra for luminescence in the ZnS, ZnS: Ag, ZnS: Cu and ZnS: Mn samples of Figure 7.18.

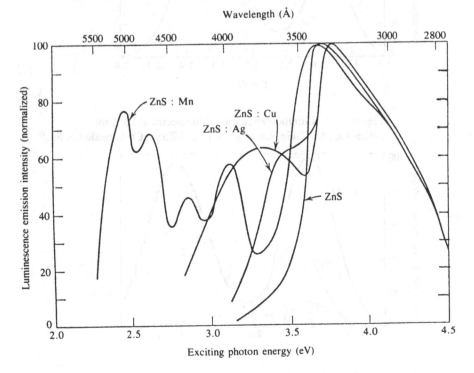

The relative magnitude of the two emission bands in ZnSe:Cu depends upon the location of the Fermi level as controlled by annealing. Figure 7.21 shows the emission spectra for ZnSe:Cu, ZnSe:Cu:Al and Se-annealed ZnSe:Cu crystals. Annealing in Se (lowering the Fermi level) increases the ratio of green to red emission, whereas incorporating Al donors (raising the Fermi level) decreases this ratio.

Figure 7.22 shows the excitation spectra of photoconductivity and both luminescence bands measured at 85 K. One transition has a

Figure 7.20. Luminescence emission spectrum for ZnSe:Cu at 77 K showing the red (1.97 eV) and green (2.34 eV) bands.[22]

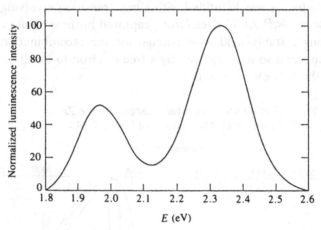

Figure 7.21. Luminescence emission spectra at 85 K for ZnSe:Cu:Al, ZnSe:Cu as grown, and ZnSe:Cu annealed in Se.[22]

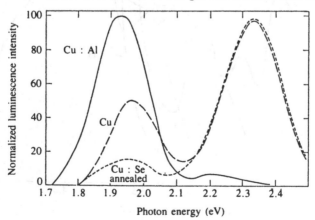

low-energy threshold of 2.3 eV and excites red luminescence and photoconductivity. The other transition has a low-energy threshold of 2.6 eV and excites green luminescence. Photo-Hall measurements showed that the photoconductivity was always n-type, and that the photo-Hall mobility was independent of photon energy over the whole extrinsic range.

Figure 7.23 shows thermal quenching of luminescence and photoconductivity. The 1.95 eV band associated with a pair transition quenches first near 50 K with an activation energy of 0.012 eV. The 2.34 eV green emission quenches at 140 K with an activation energy of 0.35 eV. The 1.97 eV red band and the photoconductivity quench simultaneously at 250 K with an activation energy of about 0.8 eV. Measurements of optical quenching of luminescence and photoconductivity also showed the same quenching spectrum for photoconductivity and the 1.97 eV red band.

Figure 7.22. Excitation spectra for photoconductivity and luminescence in ZnSe : Cu at 85 K. Curves are (triangles) photoconductivity, (circles) red luminescence, and (squares) green luminescence.

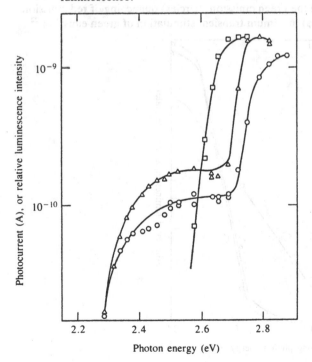

Figure 7.23. Luminescence intensity and photocurrent vs temperature in ZnSe:Cu: (circles) 1.95 eV red luminescence between 16 and 140 K, and 1.97 eV red luminescence between 216 and 320 K, (triangles) green luminescence, and (squares) photocurrent.[22]

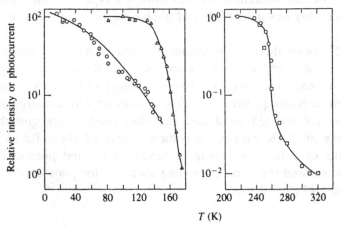

Figure 7.24. Optical quenching of red luminescence and simultaneous stimulation of green luminescence at 85 K: (squares) quenching of the green emission, (circles) quenching of red emission, and (triangles) maximum transient stimulation of green emission.[22]

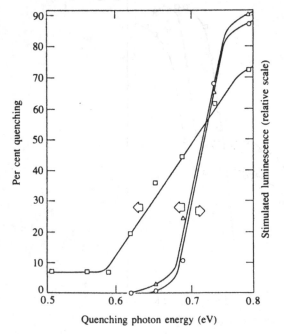

When crystals were excited at 85 K with 2.5 eV light so that the red but not the green emission was excited, optical quenching of the red emission caused a transient stimulation of the green emission. The spectra for the quenching of the red and stimulation of the green are identical, as shown in Figure 7.24. This effect can be interpreted as capture of holes by green centers after these holes are freed from red centers by optical quenching, followed by recombination of free electrons at green centers.

All of these effects (plus spectral response of optical absorption and others) lead to the description of Cu levels in ZnSe shown in Figure 7.25. The Cu centers responsible for the red and green luminescence emission may be identified as two charge states of Cu impurity substituting for Zn on the ZnSe lattice. $Cu_{Zn}'$ is responsible for the red luminescence and is located 0.72 eV above the valence band, and $Cu_{Zn}^x$ is responsible for the green luminescence and is located at 0.35 eV above the valence band. The somewhat more complicated appearance of the band diagram in Figure 7.25 results from the fact that the d orbitals of $Cu_{Zn}^x$ with the $(Ar) 3d^9$ configuration are split into $t$ and $e$ levels by the tetrahedral crystal field. The following transitions are shown in Figure 7.25.

(1)  $Cu_{Zn}^x + 0.72\,eV \rightarrow Cu_{Zn}' = h^{\cdot}$
(2)  $Cu_{Zn}^x + 0.93 \rightarrow Cu_{Zn}^{x^*}$
(3)  $Cu_{Zn}^{x^*} \rightarrow Cu_{Zn}' + h^{\cdot}$

Figure 7.25. Band diagram showing schematically $Cu_{Zn}'$, and $Cu_{Zn}^x$ centers in ZnSe:Cu. See text for identification of the numbered transitions.

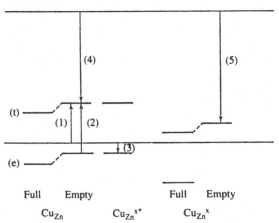

(4)  $Cu_{Zn}{}^x + e' \rightarrow Cu_{Zn}' + 1.97\,eV$ (red luminescence)

(5)  $Cu_{Zn}{}^{\cdot} + e' \rightarrow Cu_{Zn}{}^x + 2.34\,eV$ (green luminescence)

Many other such examples of the usefulness of luminescence as a technique in photoelectronic analysis could be given from the literature.

# 8
# Transient photoelectronic analysis

Rates of change are sometimes more indicative of actual kinetic processes than steady state values by themselves. The recombination process itself is characterized by a time constant that reflects the capture cross section and density of the imperfections involved. Thermally activated imperfection emptying or filling has a rate of change that depends upon the capture cross section and ionization energy. Transient methods of photoelectronic analysis focus on the rise or decay of photoinduced or electrical injection induced imperfection occupancies as they return to their thermal equilibrium values.

These methods can be applied to all of the major variables in photoelectronic behavior: photoconductivity, luminescence, and photocapacitance. From measurements of their variation with time, one can deduce values for the ionization energy, optical cross section, capture cross sections, and density of the imperfection involved.

Many processes are functions of time and of temperature. At fixed temperature their rise or decay is controlled by the values of the parameters at that temperature. Another type of phenomenon is often used, called thermally stimulated decay, in which each of these variables can again be measured as a function of time, but under conditions in which the temperature itself is varying with time, usually linearly. Thus we may consider thermally stimulated conductivity, thermally stimulated luminescence, and thermally stimulated capacitance changes.

## 8.1 Representative decay measurements

The types of transient measurements we encounter may be categorized in four major groups.

*Exponential decay rate*

The simplest decay process is one in which there is a simple exponential decay,

$$Z(t) = Z_0 \exp(-t/\tau_0) \tag{8.1}$$

with time constant of $\tau_0$, which is assumed not to be a function of time.[1] The variable $Z$ may be either photoconductivity, luminescence or capacitance. If the process involved is one in which trapped electrons are being thermally freed from traps to the conduction band, for example, then

$$1/\tau_0 = N_c \beta_n \exp(-\Delta E/kT) \tag{8.2}$$

where $N_c$ is the effective density of states in the conduction band, $\beta_n$ is the capture coefficient of the traps for electrons, and $\Delta E$ is the trap depth or ionization energy. If the decay of $Z(t)$ is measured at different temperatures, then a plot of $\ln Z(t)$ vs $t$ at each temperature is a straight line characterized by a time constant of $\tau_0(T)$. If then $\ln \tau_0(T)$ is plotted vs $1/T$ (an Arrhenius plot), a straight line is obtained with slope equal to $-\Delta E/k$ and with intercept depending on $\beta_n$. This is the simplest kind of exponential decay process.

*Strong retrapping*

In appropriate circumstances, electrons thermally freed from traps during decay may be retrapped once or many times before undergoing recombination. Such a possibility changes the shape of the decay and it is no longer a simple exponential. Many techniques seek to avoid this possibility by having a local field (as from a junction) that sweeps thermally freed carriers away and prevents retrapping.

If there is a quasi-continuous trap distribution with strong retrapping, however, a particular analysis becomes possible. Then the result of Eq. (2.33) when $n_t \gg n$ is that $n_t = G\tau_0$, where $G$ is the photoexcitation rate, $n_t$ is the density of traps that must empty for the Fermi level to drop by $kT$, and $\tau_0$ is the time required for them to empty. In the strong retrapping situation, the steady state Fermi level can be used to indicate trap occupancy, and the trap depth corresponding to the measured $n_t$ is given by $(E_c - E_t) = kT \ln(N_c/n_0)$, where $n_0$ is the free electron density (when $n \approx \Delta n$) at the beginning of the decay at $t = 0$. In this way it is possible to deduce $n_t$ vs $(E_c - E_t)$ from the decay curves.

### Rate window

A decay sampling technique particularly suited to computerized analysis is known as the 'rate window' approach. Two times are chosen during the decay process, $t_1$ and $t_2$ $(t_2 > t_1)$, and the signal $Z(t)$ at each of these times is measured. Then $[Z(t_1) - Z(t_2)]$ is plotted as a function of temperature. For low temperatures, the decay is very slow and $[Z(t_1) - Z(t_2)]$ is small, and for high temperatures, the decay is very fast and completed before $t = t_1$, so that again $[Z(t_1) - Z(t_2)]$ is small. The result is a curve exhibiting a maximum at a particular temperature, which can be suitably interpreted to obtain the trap parameters. $Z$ can represent photoconductivity, luminescence, or photocapacitance.

### Thermally stimulated behavior

A trap showing decay is subject to a time-dependent temperature, e.g., $dT = \gamma \, dt$. The effects can again be observed through conductivity, luminescence, or capacitance. A plot of the thermally stimulated signal as a function of temperature again gives rise to a curve with a maximum at a particular temperature, from which trap parameters can be derived. At temperatures below the maximum, the curve rises with increasing temperature as the rate of emptying increases: at temperatures above the maximum, the curve falls with increasing temperature as all of the traps become empty. The 'thermally stimulated' magnitude is taken as the difference between the magnitude of the signal under thermal stimulation and the value the same quantity would have had at that temperature in thermal equilibrium, e.g., $\sigma_{TSC} = \sigma_{TS} - \sigma_d$, where $\sigma_{TSC}$ is the thermally stimulated conductivity, $\sigma_{TS}$ is the conductivity measured under thermal stimulation with a particular value of $\gamma$, and $\sigma_d$ is the dark conductivity at that same temperature in thermal equilibrium. When thermally stimulated luminescence (sometimes misnamed 'thermoluminescence') is being observed, the effect of a sudden emission of light in a dark room as the temperature is raised can have a dramatic aesthetic effect, and such measurements were named 'glow curves' many years ago.

## 8.2    Decay of a single level

First we consider the simple case of a single trap depth, with thermal emptying of the trap followed either by retrapping or recombination. There are two cases: ($a$) constant lifetime $\tau$ during the decay, which means that the density of empty recombination centers

$(N_R - n_R)$ does not change appreciably during the decay, and (b) varying lifetime during the decay, the extreme case of which corresponds to $N_R - n_R = n_t + n \approx n_t$, where $n_t$ is the density of electron occupied traps and we are using the $n \approx \Delta n$ 'insulator' approximation. Situation (b) corresponds to the case where there is a hole in a recombination center for each electron in the trap.

The general equation describing the decay is

$$dn/dt = -n/\tau - dn_t/dt \qquad (8.3)$$

with

$$\tau = 1/[\beta_n(N_R - n_R)] \qquad (8.4)$$

and

$$dn_t/dt = n\beta_t(N_t - n_t) - n_t N_c \beta_t \exp[-(E_c - E_t)/kT] \qquad (8.5)$$

The first term on the right of Eq. (8.5) represents the retrapping of electrons by empty traps, and the second term represents the thermal excitation of trapped electrons to the conduction band from a trap with electron capture coefficient $\beta_t$ and trap depth $(E_c - E_t)$. Using our previously introduced notation of $P_t = N_c \beta_t \exp[-(E_c - E_t)/kT]$, Eq. (8.5) becomes

$$dn_t/dt = n\beta_t(N_t - n_t) - n_t P_t \qquad (8.6)$$

Two crucial questions arise: (a) Is $\tau$ constant with time? (b) Is retrapping important? The answer to the second question will depend to some extent at least on the magnitude of $\beta_t$.

### $\tau$ constant; no retrapping; neglect $dn/dt$

The simplest way to proceed would be to assume that the rates that control Eq. (8.3) are $n/\tau$ and $dn_t/dt$, i.e., once an electron is thermally freed, it recombines virtually simultaneously. We have therefore that $n \approx -\tau \, dn_t/dt$, and since $dn_t/dt = -n_t P_t$,

$$n_t = n_{t0} \exp(-P_t t) \qquad (8.7)$$

and

$$dn_t/dt = -n_{t0} P_t \exp(-P_t t) \qquad (8.8)$$

We conclude that

$$n \approx n_{t0} P_t \tau \exp(-P_t t) \qquad (8.9)$$

Measurements of $\ln n$ vs $t$ give straight lines at different temperatures with slopes corresponding to $P_t$. A plot of $\ln(P_t T^{-3/2})$ vs $1/T$ gives a straight line with slope equal to $-(E_c - E_t)/k$ and intercept proportional to $\beta_t$. This formulation describes only the contribution to the decay from trapped electrons and does not include a contribution from free electrons.

*$\tau$ not constant; no retrapping; neglect $dn/dt$*

The situation is radically altered if the lifetime is not constant. If, in fact, $\tau = 1/\beta_n n_t = \exp(P_t t)/n_{t0}\beta_n$, then

$$n \approx P_t/\beta_n \qquad (8.10)$$

and in fact we cannot describe decay in this approximation at all since the increase of lifetime with time just compensates for the decay itself.

*$\tau$ constant; no retrapping; include $dn/dt$*

We need not make the extreme assumption above involved in neglecting $dn/dt$, but do a more complete calculation including $dn/dt$ and the contribution of free electrons to the decay. Eq. (8.3) becomes

$$dn/dt + n/\tau = n_{t0}P_t \exp(-P_t t) \qquad (8.11)$$

If $n = n_L$ at $t = 0$, the solution of Eq. (8.11) is

$$n = n_{t0}P_t\tau \exp(-P_t t)/(1 - P_t\tau)$$
$$+ [n_L - n_{t0}P_t\tau/(1 - P_t\tau)]\exp(-t/\tau) \qquad (8.12)$$

If $P_t \ll 1/\tau$, then Eq. (8.12) reduces to Eq. (8.9). The first term on the right of Eq. (8.12) describes the contribution of traps to the decay, and the second term describes the contribution of free electrons to the decay with a time constant equal to the free electron lifetime.

## 8.3 Strong retrapping

The assumption of strong retrapping implies a strong interaction between trap emptying and filling, i.e., a situation where occupancies can be described in terms of the steady state Fermi level. Because of this assumption, it becomes possible to relate $dn_t/dt$ directly to $dn/dt$. Expressions for $n_t$ and $n$ are

$$n_t = N_t/((1/2) \exp\{[(E_c - E_{Fn}) - (E_c - E_t)]/kT\} + 1) \qquad (8.13)$$

and

$$n = N_c \exp[-(E_c - E_{Fn})/kT] \qquad (8.14)$$

*Trap above Fermi level: $E_t > E_{Fn}$*

In this case we can neglect the one in the denominator of Eq. (8.13) and

$$n_t = 2(N_t/N_c)n\{\exp[(E_c - E_t)/kT]\} \tag{8.15}$$

so that

$$\frac{dn_t}{dt} = 2(N_t/N_c)\{\exp[(E_c - E_t)/kT]\}\frac{dn}{dt} \tag{8.16}$$

If Eq. (8.16) is substituted into Eq. (8.3), we obtain

$$dn/dt = -n/(\tau\{1 + 2(N_t/N_c)\exp[(E_c - E_t)/kT]\}) \tag{8.17}$$

which gives

$$n = n_L \exp(-t/\tau^*) \tag{8.18}$$

with

$$\tau^* = \tau\{1 + 2(N_t/N_c)\exp[(E_c - E_t)/kT]\} \tag{8.19}$$

In this case the decay even with strong retrapping is exponential, but with a time constant that depends on trap emptying as well as free carrier decay.

Consideration of the special case of $E_t = E_{Fn}$ following this same procedure shows that the same type of decay occurs as for the case of $E_t > E_{Fn}$ in Eq. (8.18), but that the corresponding $\tau^*$ is slightly altered to

$$\tau^* = \tau\{1 + (\tfrac{2}{9})(N_t/N_c)\exp[(E_c - E_t)/kT]\} \tag{8.20}$$

*Trap below Fermi level: $E_t \ll E_{Fn}$*

For this condition $n \gg N_c\exp[-(E_c - E_t)/kT]$ and trap emptying is very slow. Eq. (8.3) becomes

$$dn/dt = -n/\tau - \{N_cN_t\exp[-(E_c - E_t)/kT]/2n^2\}\,dn/dt \tag{8.21}$$

with solution (after inverting to obtain $dt/dn$):

$$\tau\ln(n_L/n) + \tfrac{1}{4}\tau N_cN_t\exp[-(E_c - E_t)/kT](1/n^2 - 1/n_L^2) = t \tag{8.22}$$

The first term on the left corresponds to the free carrier decay, and if taken by itself gives $n = n_L\exp(-t/\tau)$. The second term on the left corresponds to trapped carrier decay, and if taken by itself gives

$$(n/n_L)^2 = (n_L^2 t\{\tfrac{1}{4}\tau N_cN_t\exp[-(E_c - E_t)/kT]\} + 1)^{-1} \tag{8.23}$$

which is certainly not an exponential.

$\tau$ *not constant*: $E_t > E_{Fn}$

If $\tau$ is not constant, then the form of the decay may be completely changed from the exponential decay that characterized this case with constant lifetime. If, for example, $\tau = 1/\beta_n n_t = N_c \exp[-(E_c - E_t)/kT]/\beta_n N_t n$, we get

$$dn/dt = -n^2/A \qquad (8.24)$$

with

$$A = (N_c/\beta_n N_t) \exp[-(E_c - E_t)/kT]\{1 + (N_t/N_c) \exp[(E_c - E_t)/kT]\}.$$

with solution of the form

$$n = n_L A/(A + n_L t) \qquad (8.25)$$

This is an example of a typical hyperbolic decay with $n \propto t^{-1}$ at long times.

### Trap analysis using a strong retrapping perspective

In Section 8.1 we referred to the possibility of proceeding from measured decay curves to inferred trap distributions on the basis of a strong retrapping perspective. We make use of the steady state Fermi level location to describe the occupancies during decay because the assumption of strong retrapping implies comparable rates of thermal emptying and retrapping.

Start with the measurement under photoexcitation, so that we have $n_L = G\tau$, which enables us to determine $\tau$. Then measure the time $\tau_0$ it takes for $n$ to decrease to $n_L/e$, and calculate $n_t = n_L(\tau_0/\tau)$ and $E_c - E_{Fn} = kT \ln(N_c/n_L)$. We then conclude that there are $n_t/kT$ traps per unit volume per eV with depth corresponding to $(E_c - E_{Fn})$. Repeat with other light intensities at other temperatures for a complete mapping of the trap distribution.

### Intermediate retrapping

It is possible to describe an intermediate situation between no and strong retrapping. For the case of constant $\tau$ and neglecting $dn/dt$ in Eq. (8.3), we obtain

$$dn_t/dt = -\epsilon n_t P_t = -\{\beta_n(N_I - n_I)/[\beta_n(N_I - n_I) + \beta_t(N_t - n_t)]\}n_t P_t \qquad (8.26)$$

where $\epsilon$ represents the net fraction of traps emptied. Integration gives

$$[\beta_n(N_I - n_I) + \beta_t N_t] \ln(n_t/n_{t0}) - \beta_t(n_t - n_{t0}) = -\beta_n(N_I - n_I)P_t t \qquad (8.27)$$

If the second term on the left is neglected, then

$$n_t = n_{t0} \exp(-\alpha P_t t) \qquad (8.28)$$

with

$$\alpha = \{\beta_n(N_I - n_I)/[\beta_n(N_I - n_I) + \beta_t N_t]\} \qquad (8.29)$$

so that

$$n \approx -(dn_t/dt)\tau = n_{t0}\alpha P_t \tau \exp(-\alpha P_t t) \qquad (8.30)$$

Once again the form of the decay is completely changed if $\tau$ is not constant. For example, if $n_t = (N_I - n_I)$, $n_{t0} = N_t$, $\tau = 1/\beta_n n_t$,

$$dn_t/dt = -\beta_n n_t^2 P_t/[\beta_t N_t + (\beta_n - \beta_t)n_t] \qquad (8.31)$$

giving

$$(\beta_t/\beta_n)N_t[1/n_t - 1/N_t] - [(\beta_n - \beta_t)/\beta_n]\ln(n_t/N_t) = P_t t \qquad (8.32)$$

If, for simplicity, we set $\beta_n = \beta_t$, $n_t = N_t/(1 + P_t t)$, and since $n \approx -(dn_t/dt)(1/\beta_n n_t)$, we again arrive at a power law form for $n$:

$$n \approx P_t/[\beta_n(1 + P_t t)] \qquad (8.33)$$

### Rise curves

The form of the rise curve has already been given in Eq. (2.32) for the case of no trapping. A similar but slightly modified result is obtained for the case of strong retrapping described in Eq. (8.16) for $E_t > E_{Fn}$. We can write

$$dn_t/dt = 2(N_t/N_c) \exp[-(E_c - E_t)/kT] \, dn/dt = \theta \, dn/dt \qquad (8.34)$$

so that

$$dn/dt = G - n/\tau - \theta \, dn/dt \qquad (8.35)$$

with solution

$$n = G\tau\{1 - \exp[-t/(1 + \theta)\tau]\} \qquad (8.36)$$

so that the effective response time $\tau_0 = (1 + \theta)\tau$.

## 8.4    Specific trap distributions

We consider next a few specific trap distributions and investigate the transient behavior associated with them.

### Discrete set of traps

If strong retrapping can be assumed, then the decay curves associated with a discrete set of traps can be calculated. Assume that there is a discrete set of traps with depth $(E_c - E_{ti})$ and density $N_{ti}$ associated with the $i$th trap, a major recombination center with density $N_R$, and that the lifetime $\tau = 1/\beta_R(N_R - n_R)$ can be considered effectively constant over the range of the decay measurements. The calculation proceeds in the following steps:

(1) Calculate $E_{FnL}$ in the light from $n_L = N_c \exp[-(E_c - E_{FnL})/kT]$.

(2) Calculate $\tau$ from $n_L = G\tau$.

(3) Calculate the trap occupancies in the light from

$$n_{tiL} = N_{ti}/\{\tfrac{1}{2}\exp[-(E_{ti} - E_{FnL})/kT] + 1\} \qquad (8.37)$$

(4) Increase $E_{FnL}$ by $kT$ so that $E_{Fn1} = E_{FnL} + kT$, corresponding to $n_1 = n_L/e$.

(5) Calculate $n_{t1} = \sum (n_{tiL} - n_{ti1})$.

(6) Calculate $\tau_{01} = \tau(1 + n_{t1}/n_L)$.

(7) Repeat the calculation with stepwise increases of $E_{Fn}$ by $kT$.

(8) Plot $n$ vs $t$.

Examples of the results of such calculations are given in Figure 8.1. The discrete set of traps assumed has been reported for CdS,[2] with depth and density parameters given in the figure caption. Figure 8.1($a$) compares decay curves on a semilog electron density vs time scale at 100 K for the extreme case of no retrapping in which we simply sum independent exponential decays for each of the seven traps with the case of strong retrapping. Figure 8.1($b$) shows the decay curves that are calculated at two temperatures assuming strong retrapping. Figure 8.1($c$) takes the curves of Figure 8.1($b$) as given experimental data, and then analyzes them to obtain the corresponding trap distribution. The calculation process in Figure 8.1($c$) gives an artificial width to each trap depth that is more pronounced at higher temperatures, but the indicated trap depths agree well with those assumed for the calculation of Figure 8.1($b$).

### Continuous uniform trap distribution

If strong retrapping can be assumed, then the decay curves resulting from a continuous uniform trap distribution, $N_t(E)$, can be calculated. When the Fermi level has moved from $E_{Fn0}$, its equilibrium value, to $E_{Fn}$ under illumination, the density of trapped electrons is

$$n_t = N_t(E)(E_{Fn} - F_{Fn0}) = kT\,N_t(E)\ln(n/n_0) \qquad (8.38)$$

Therefore

$$dn_t/dt = (dn_t/dn)(dn/dt) = N_t(E)(kT/n)\,dn/dt \qquad (8.39)$$

The decay curve can therefore be calculated, assuming constant $\tau$,

$$\mathrm{d}n/\mathrm{d}t = -n/\tau - N_t(E)(kT/n)\,\mathrm{d}n/\mathrm{d}t \qquad (8.40)$$

with solution

$$\tau \ln(n_L/n) + \tau kT N_t(E)(1/n - 1/n_L) = t \qquad (8.41)$$

If the first term dominates, the decay of free carriers proceeds as usual according to $n = n_L \exp(-t/\tau)$, whereas if the second term due to traps dominates, the decay is hyperbolic at long times and is given by

$$n = n_L \tau^*/(\tau^* + n_L t) \qquad (8.42)$$

with $\tau^* = \tau kT N_t(E)$.

Figure 8.1. Calculated curves assuming a particular set of electron traps reported for CdS assuming constant lifetime over the range shown.[2] Seven discrete trap depths (eV) and densities (cm$^{-3}$) are as follows: 0.16, $2 \times 10^{14}$; 0.31, $1.9 \times 10^{14}$; 0.38, $1.1 \times 10^{15}$; 0.42, $2.4 \times 10^{14}$; 0.53, $1.8 \times 10^{15}$; 0.69, $1.2 \times 10^{15}$; 0.96, $9.7 \times 10^{14}$. (*a*) Calculated decay curves. Electron density vs $t$ at 100 K; no retrapping (solid curve), strong retrapping (dashed curve). (*b*) Calculated decay curves. Electron density vs $t$ at 100 K (open circles) and 300 K (filled circles) for strong retrapping. (*c*) Using the decay curves of (*b*) to calculate trap density vs trap depth assuming strong retrapping, (open circles) 300 K, (triangles) 100 K.

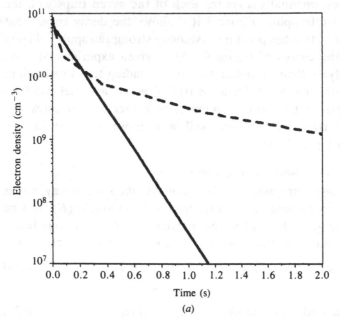

*(a)*

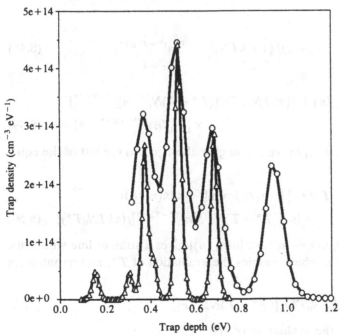

Figure 8.1. (*Continued*)

*(b)*

Figure 8.1. (*Continued*)

*(c)*

The rise curve for this distribution can be calculated from

$$dn/dt = G - n/\tau - (kTN_t(E)/n)\,dn/dt \qquad (8.43)$$

If $n \gg kTN_t(E)$, free carrier behavior dominates and $n = G\tau[1 - \exp(-t/\tau)]$ as usual. If $n \ll kTN_t(E)$, traps dominate and

$$n = G\tau/\{1 + [(G\tau - n_0)/n_0]\exp(-t/\tau_0)\} \qquad (8.44)$$

### Continuous exponential trap distribution

Again assuming strong retrapping and following the same procedure as for the uniform trap distribution, calculate

$$n_t = \int N_0 \exp[-(E_c - E)/kT^*]\,dE \qquad (8.45)$$

where $T^*$ is a characteristic parameter defining the exponential trap distribution. The lower limit is $(E_c - E_{Fn})$ and the upper limit is $(E_c - E_{Fn0})$. The result of integrating Eq. (8.45) can be written as

$$n_t = kT^* N_0 N_c^{-T/T^*}(n^{T/T^*} - n_0^{T/T^*}) \qquad (8.46)$$

Therefore

$$dn_t/dt = kTN_0 N_c^{-T/T^*} n^{(T - T^*)/T^*}\,dn/dt \qquad (8.47)$$

giving

$$dn/dt = -(n/\tau)/\{1 + kTN_0 N_c^{-T/T^*} n^{(T - T^*)/T^*}\} \qquad (8.48)$$

with solution

$$\tau \ln(n_L/n) + \{(\tau kTN_0 T^*)/[(T^* - T)N_c^{T/T^*} n_L^{\{T^* - T)/T^*\}}]\}$$
$$\times [(n_L/n)^{(T^* - T)/T^*} - 1] = t \quad (8.49)$$

For large enough $n_L/n$ we can neglect the 1 just to the left of the equal sign, and obtain

$$[(T^* - T)/T^*]\ln(n_L/n) = \ln[t - \tau\ln(n_L/n)]$$
$$+ \ln\{[(T^* - T)N_c^{T/T^*} n_L^{(T^* - T)/T^*}]/(\tau kTN_0 T^*)\} \quad (8.50)$$

A plot of $\ln(n_L/n)$ vs $\ln[t - \tau\ln(n_L/n)]$ gives a straight line with slope of $T^*/(T^* - T)$, which enables the evaluation of $T^*$, and an intercept $t_i$ given by

$$t_i = (\tau kTN_0 T^*)/[(T^* - T)N_c^{T/T^*} n_L^{(T^* - T)/T^*}] \qquad (8.51)$$

which enables the evaluation of $N_0$.

Now it is also true that $\tau_0 = \tau(1 + n_t'/n)$ where

$$n_t' = \int N_0 \exp[(-E_c - E)/kT^*]\,dE \qquad (8.52)$$

where the lower limit is $(E_c - E_{Fn})$ and the upper limit is $(E_c - E_{Fn} + kT)$. The result of the integration is

$$n_t' = kT^* N_0 n^{T/T^*} N_c^{-T/T^*}[1 - \exp(-T/T^*)] \qquad (8.53)$$

which then gives

$$\tau_0 = \tau + [(T^* - T)/T][1 - \exp(-T/T^*)]t_i \qquad (8.54)$$

and provides an alternative way of calculating $\tau_0$ for comparison.

An example of the application of this model is provided by decay measurements in CdSSe sintered layers,[3] providing the kind of imperfect system in which one might expect to find an exponential trap distribution.[4] Photoconductivity decay curves were measured from $10^{-4}$ to $10^{-1}$ s at three different temperatures near room temperature, and for two different photoexcitation intensities.

In one approach to the data each of the decay curves was analyzed by plotting $\ln(n_L/n)$ vs $\ln[t - \tau \ln(n_L/n)]$ for values of $(n_L/n) > 20$ in every case and $>100$ in most cases. Values of $T^*$ between 690 and 937 K were obtained from the slope; and values of $N_0$ between $1.1 \times 10^{17}$ and $7.4 \times 10^{17}$ cm$^{-3}$ were obtained from the intercept $t_i$ according to Eq. (8.51). Values of $\tau_0$ calculated from Eq. (8.54) were also compared with measured values, and it was found that on the average calculated values of $\tau_0$ were about 0.7 times measured values of $\tau_0$, providing reasonably good agreement.

In a second approach to the data, all of the experimental decay curves were analysed together using Fermi level analysis according to $n_t' = (\tau_0'/\tau - 1)n$, as described in Section 8.3, to obtain a single trap distribution. The results are shown in Figure 8.2(a), showing that the data are consistent with an exponential trap distribution between 0.3 and 0.6 eV with $N_0 = 2 \times 10^{17}$ cm$^{-3}$ and $T^* = 852$ K. In the region between 0.42 and 0.46 eV, for example, a point on the trap distribution curve is obtained from each of the six temperature and excitation intensity pairs; the average deviation of these six points from the curve is 25%.

Once this trap distribution has been obtained by analysis of the decay curves, it can in turn be used to reconstruct each of the separate decay curves without any adjustable parameters. The result is given in Figure 8.2(b). Overall good agreement is found between the model of

Figure 8.2. (*a*) The calculated variation of trap density as a function of $(E_c - E_{Fn})$ as a result of Fermi level analysis of measured decay curves for a sintered CdSSe layer assuming strong retrapping.[3] Measurements were made at three different temperatures for two different light intensities; the meaning of the symbols is given in the insert. (*b*) Comparison of experimental decay data, indicated by the points with the same symbol code as in (*a*) (plus data points for 9.4 $\mu$W at 298 K indicated by circles with a horizontal crossbar), with theoretical curves drawn for $N_0 = 2 \times 10^{17}$ cm$^{-3}$ and $T^* = 852$ K. The theoretical curves are as follows: (1) infinite excitation intensity, for the 298 K lifetime, (2) 940 $\mu$W, 348 K, (3) 940 $\mu$W, 323 K, (4) by coincidence, both 940 $\mu$W at 298 K, and 94 $\mu$W at 348 K, (5) 94 $\mu$W, 323 K, (6) 94 $\mu$W, 298 K, and (7) 9.4 $\mu$W, 298 K.

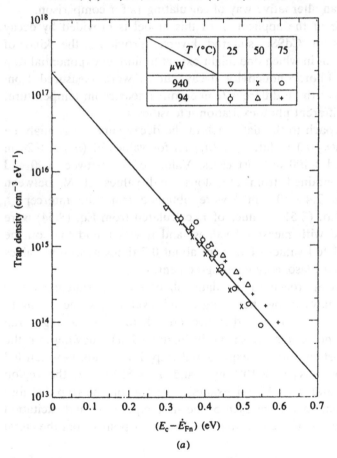

| $T$ (°C) | 25 | 50 | 75 |
|---|---|---|---|
| $\mu$W | | | |
| 940 | ▽ | x | o |
| 94 | ◇ | △ | + |

(*a*)

a continuous exponential trap distribution with strong retrapping and the experimental data.

## 8.5 Thermally stimulated luminescence

Examples of thermally stimulated luminescence – 'glow curves' – have been observed for literally hundreds of years, but have been used analytically only in the last half century.[5] In a typical case, a material is cooled down (usually in the dark to 77 K), illuminated by radiation capable of creating free and hence trapped carriers, returned to the dark, and then heated (usually at a linear rate) while the emission intensity of a particular luminescence emission band is measured.

A wide variety of basic conclusions can be drawn without detailed

Figure 8.2. (*Continued*)

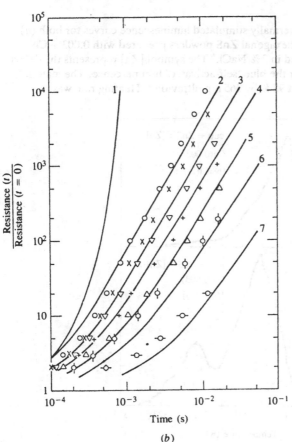

(*b*)

quantitative analysis of such curves.[6] An illustration is given in Figure 8.3 for two powder samples of ZnS with 0.003% Cu, one with the cubic structure and the other with the hexagonal structure, measured at a very slow rate of heating such that resolution of individual subpeaks becomes possible.

Figure 8.4 shows a set of glow curves for the hexagonal samples of ZnS:Cu with a range of Cu concentrations from 0 to 0.3%. These curves show that (1) a large fraction of the traps found in ZnS:Cu are not directly associated with Cu, (2) the incorporation of Cu up to a critical proportion causes the appearance of high-temperature thermally stimulated luminescence, (3) the incorporation of Cu in concentrations higher than this critical proportion leads to a disappearance of high-temperature thermally stimulated luminescence.

An important factor that must be taken into account is the temperature dependence of the luminescence efficiency, since if

Figure 8.3. Thermally stimulated luminescence curves for both (*a*) cubic and (*b*) hexagonal ZnS powders prepared with 0.003% Cu impurity heated in 2% NaCl.[6] The symbol [Zn] represents the defect responsible for the blue 'self-activated' luminescence. The materials were excited at 98 K by 365 nm ultraviolet. Heating rate was 0.033 K s$^{-1}$.

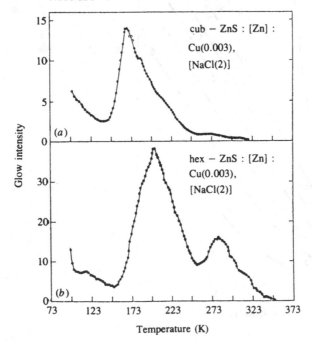

recombination occurs non-radiatively, the corresponding thermally stimulated luminescence will disappear from the glow curve. This might be mistakenly attributed to a decrease in the density of a particular trap or set of traps if the temperature dependence of the luminescence efficiency were not considered. In the case of the data shown in Figure 8.4, the temperature dependence for the green emission due to Cu is given in Figure 8.5. Comparison of the two figures shows that the disappearance of the high-temperature glow curve peak with increasing Cu above 0.03% is due to a loss of high-temperature luminescence efficiency with increasing Cu.

The apparent fine-structure in the glow curve of Figure 8.4 suggests

Figure 8.4. Thermally stimulated luminescence curves for hexagonal ZnS powders with Cu concentrations varying from zero to 0.3%. The materials were excited at 77 K by 365 nm ultraviolet. Heating rate was 0.33 K s$^{-1}$.

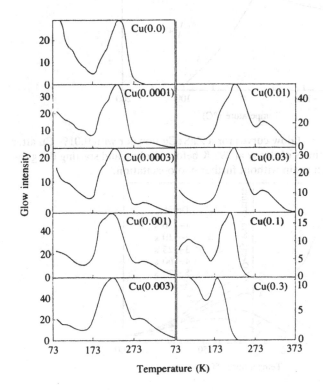

Figure 8.5.  Green luminescence intensity due to Cu impurity in ZnS powders as a function of temperature in (*a*) cubic and (*b*) hexagonal ZnS, for several different Cu concentrations.[7]

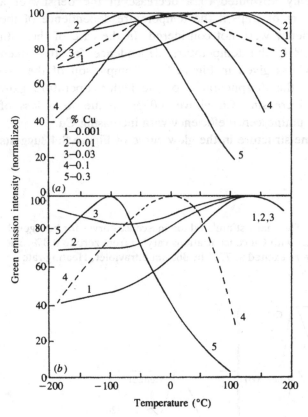

% Cu
1–0.001
2–0.01
3–0.03
4–0.1
5–0.3

Figure 8.6.  Glow curves for a ZnS:Cu powder with 0.01% Cu after various times of decay at 292 K before cooling and starting the glow curve run again without further photoexcitation.[6]

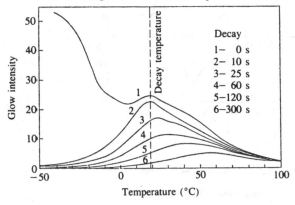

Decay
1– 0 s
2– 10 s
3– 25 s
4– 60 s
5–120 s
6–300 s

that there are not just three trap depths corresponding to the main structure of the curves, but many more. One way to test this is to measure 'decayed' glow curves after several different times of decay at a fixed temperature. For example, Figure 8.6 shows the glow curves for a ZnS:Cu powder after heating to 292 K, holding the temperature constant for several lengths of time up to 300 s, recooling, and then reheating without further photoexcitation for the measurement of the remaining glow curve.[6] If only a single trap depth were involved, the maximum of the glow curve should occur at the same temperature regardless of time of decay; the fact that this is not the case confirms the speculation that there are several different trap depths involved in this high temperature glow curve peak.

Another process that can be demonstrated through the use of glow curves is retrapping, as shown in Figure 8.7.[6] The case cited here is not for retrapping by traps of the same depth as are emptying, as discussed in the previous discussion of decay curves in Section 8.3, but rather retrapping of electrons freed from shallow traps by empty deep traps. Thus the extent of retrapping depends upon the fraction of deep traps that are full at the beginning of the glow curve measurement. Figure 8.7 shows the shape of the glow curve for different times of excitation ranging from 2425 s, when it may be assumed that all traps, both shallow and deep, are filled, to only 12 s, when almost all of the electrons freed from shallow traps are retrapped in empty deeper traps.

## 8.6    Thermally stimulated conductivity

Analysis of thermally stimulated conductivity (TSC) data allows an estimate of both the trap density and depth, as well as other parameters depending on the suitability of the model chosen. If the thermally stimulated current density $J_{TSC}$ is defined as

$$J_{TSC} = J_{TS} - J_D \tag{8.55}$$

where $J_D$ is the dark current measured at a given temperature in thermal equilibrium, and $J_{TS}$ is the thermally stimulated current at that temperature after photoexcitation at a low temperature, then the density of a particular trap can be estimated from the area under the TSC curve corresponding to that trap:

$$N_t = \int J_{TSC} \, dt / qVG^*) \tag{8.56}$$

where $V$ is the sample volume, and $G^*$ is an attempt to approximate the relevant gain, i.e., the number of charges measured in the external

Figure 8.7. Investigation of retrapping by means of glow curves for a
ZnS:Cu luminescent powder. Curves in (*a*)–(*c*) are obtained after
excitation at constant intensity but for times varying from 2425 to
12 s; (*d*) shows the dependence of the ratio of the area under the
low-temperature peak to the area under the high-temperature peak
on the excitation time; (*e*) shows the proportion of electrons freed
from shallow traps which are retrapped by deep traps, as a function
of the proportion of the deep traps that were empty.[6]

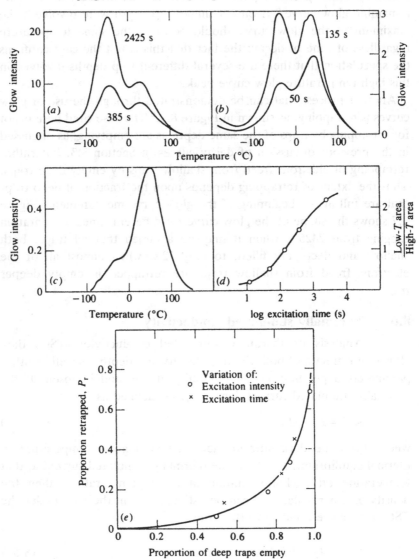

circuit per electron freed from a trap. A reasonable approximation for $G^*$ may be obtained by measuring the photon flux $F$ needed to give a photocurrent equal to $J_{TSC}$ at the temperature $T_{max}$ of the TSC maximum,

$$G^* = (\Delta J = J_{TSC})/qF \tag{8.57}$$

This assumes that the steady state Fermi level has the same location under TSC conditions as under photoconductivity conditions, and this may introduce some error.

In order to obtain mathematical descriptions of TSC curves, several possible models must be considered.

*Discrete trap level; no retrapping*

We start with Eq. (8.9)

$$n = n_{t0}\tau P_t \exp(-P_t t) \tag{8.9}$$

with $P_t = N_c\beta_t \exp[-(E_c - E_t)/kT]$. Now we let $dT = \gamma\, dt$, and integrate

$$n_{TSC} = (n_{t0}\tau P_t) \exp\left[-(1/\gamma) \int P_t\, dT\right] \tag{8.58}$$

The term in brackets before the exponential describes the variation of $n_{TSC}$ with $T$ at low $T$ when the traps are still almost full. Over this range $n_{TSC} \propto \exp[-(E_c - E_t)/kT]$ and a plot of $\ln n_{TSC}$ vs $1/T$ gives a straight line with slope of $-(E_c - E_t)/kT$; this is the method of 'decayed TSC' for the determination of trap depth. To determine the value of $n_{TSC}$ at the maximum, we need to set $d(\ln n_{TSC})/dT = 0$. Since

$$\ln n_{TSC} = \ln(n_{t0}\tau N_c\beta_t) - (E_{ct} - E_t)/kT$$

$$- (1/\gamma) \int N_c\beta_t \exp[-(E_c - E_t)/kT]\, dT \tag{8.59}$$

the result is that

$$E_c - E_t = kT_m \ln[kT_m{}^2 N_c\beta_t/\gamma(E_c - E_t)] \tag{8.60}$$

where $T_m$ is the temperature at the maximum value of $n_{TSC}$. Eq. (8.60) is a transcendental equation that must be solved iteratively.

An alternative approach is to plot $\ln(T_m{}^2/\gamma)$ vs $1/T_m$ for several different heating rates. The slope of the straight line obtained gives $(E_c - E_t)$ and the intercept allows the calculation of $\beta_t$. The disadvantage is that $T_m$ does not vary rapidly with $\gamma$, and the accuracy of the results is therefore limited.

Figure 8.8 shows typical TSC curves measured for five different heating rates varying by a factor of 20 between slowest and fastest. The total shift in $T_m$ for the larger, high-temperature TSC peak is about 45°, which is about 20% of the value of $T_m$ at the lowest heating rate.

### Discrete trap level; strong retrapping

It is convenient to rewrite the problem in this case in terms of the total density of electrons $n_T = n + n_t$. Then

$$dn_T/dt = -n/\tau = -n_T/\tau + n_t/\tau = -(n_T/\tau)[1 - n_t/(n + n_t)]$$

$$\approx -(n_T/\tau)(n/n_t) = -(n_T/\tau)(N_c/N_t) \exp[-(E_c - E_t)/kT] \quad (8.61)$$

which yields

$$n_T = n_{T0} \exp[-(1/\gamma)] \int (N_c/N_t\tau) \exp[-(E_c - E_t)/kT] \, dT$$

$$(8.62)$$

Figure 8.8. Thermally stimulated current curves for a CdS crystal, measured at five different heating rates from 0.024 to 0.49 °C s$^{-1}$.

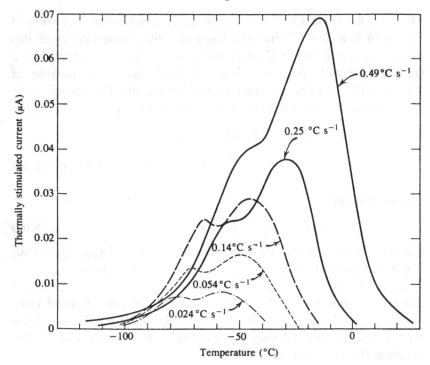

and consequently

$$n_{TSC} = n_{T0}(N_c/N_t) \exp[-(E_c - E_t)/kT]$$

$$\times \exp[-(1/\gamma)] \int (N_c/N_t\tau) \exp[-(E_c - E_t)/kT] \, dT \quad (8.63)$$

Once again, for small $t$, i.e., for low temperatures, $n_{TSC} \propto \exp[-(E_c - E_t)/kT]$, providing the same opportunity described above to determine the trap depth by measuring the initial slope of a $\ln n_{TSC}$ vs $1/T$ plot in a 'decayed TSC' experiment.

To calculate the maximum, we set $d(\ln n_{TSC})/dT = 0$, which yields

$$E_c - E_t = kT_m \ln(N_c k T_m^2/N_t \gamma \tau E_t) \quad (8.64)$$

This is similar to Eq. (8.60) for the case of no retrapping, but in this case the result is independent of $\beta_t$. Once again, however, a plot of $\ln(T_m^2/\gamma)$ vs $1/T_m$ yields a straight line with slope $(E_c - E_t)$.

No retrapping and strong retrapping cases have the following three features in common: (1) in both cases, $E_c - E_t = AkT_m$, with values of $A$ typically between 20 and 25,[8] (2) in both cases a 'decayed TSC' measurement can be made to determine the trap depth $(E_c - E_t)$, and (3) in both cases a plot of $\ln(T_m^2/\gamma)$ vs $1/T_m$ yields a straight line with slope $(E_c - E_t)$. The no retrapping case allows one in principle to be able to determine $\beta_t$, whereas the strong retrapping case does not.

If strong retrapping is believed to be present, a simple alternative to the above analysis is often sufficiently accurate. It is simply assumed that the Fermi level lies at the trap level when the TSC is a maximum.

$$E_c - E_t = E_{Fn}|_{T=T_m} = kT_m \ln(N_c/n_m) \quad (8.65)$$

Typically the trap depth determined from this simple Fermi level calculation is within $kT_m$ (i.e., about 5%) of the trap depth determined from the more detailed methods described above.

### 'Decayed TSC' measurement

An example of a TSC measurement and its 'decayed TSC' counterpart is given for a high-resistivity Zn-doped Si crystal in Figures 8.9 and 8.10.[9] Figure 8.9 shows the normal TSC curve, together with a corrected curve taking into account measured changes in electron life (hence gain $G^*$) with temperature. As in the case of thermally stimulated luminescence, it is important to take into account the variation of luminescence efficiency with temperature, so in measurements of thermally stimulated conductivity, it is important to take variations in free carrier lifetime with temperature into account.

Figure 8.10 shows the decayed TSC curves for several decay temperatures. The slope of the decayed TSC curve increases with the decay temperature, as deeper traps are being sampled. It appears that there are three major trap depths in this sample. The slope for different decay temperatures is constant at low temperatures at 0.17 eV, at intermediate temperatures at 0.26 eV, and the highest temperature slope corresponds to 0.42 eV.

### Hall effect during thermal stimulation

Just as there are cases in which the dominant carrier in photoconductivity may change with temperature, so also there are cases where the TSC may be due to electron traps in one temperature range and hole traps in another. The understanding of these effects is not possible unless a technique is used that enables one to determine the sign of the charge carriers during thermal stimulation. Clearly simultaneous measurement of the Hall effect is such a technique.

Figure 8.9. TSC for a Si:Zn crystal.[9] The solid curve gives the conductance actually measured, and the dashed curve gives the approximate trap distribution shape derived from applying corrections accounting for lifetime variation with temperature to the data.

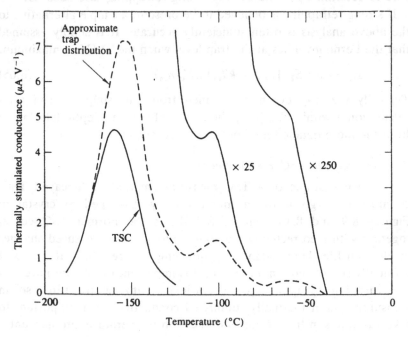

Figure 8.11 shows two examples of high-resistivity GaAs crystals (the same as those involved in Figure 7.7(*b*)) for which the Hall effect was measured during the measurement of thermally stimulated conductivity.[10] The first traps to empty at the lowest temperature are hole traps. Then the Hall mobility drops at about the center of the TSC peak to a minimum, more pronounced in Figure 8.11(*a*) than in Figure 8.11(*b*). This drop may be associated with the emptying of an electron trap with slightly larger ionization energy than the hole trap giving rise to a two-carrier Hall effect regime. After the principal low-temperature peak has been passed, there is an indication of

Figure 8.10. Decayed TSC curves for the Si:Zn crystal of Figure 8.9.[9]

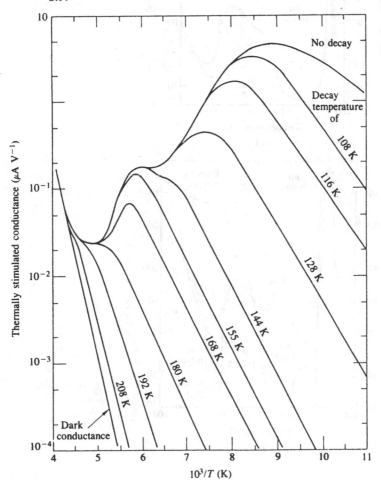

Figure 8.11. Temperature dependence of TSC and Hall mobility measured simultaneously with the TSC for (*a*) an annealed float-zone crystal of GaAs and (*b*) an annealed Bridgman grown crystal of GaAs.[10]

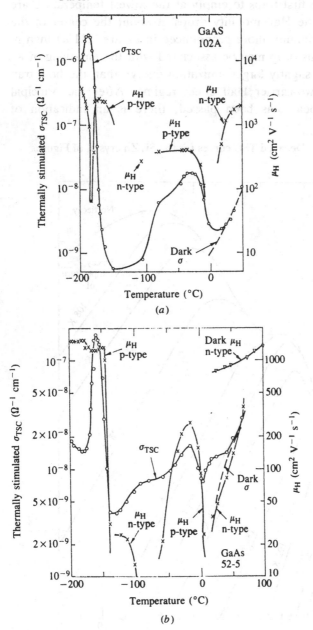

(*a*)

(*b*)

electron trap emptying at the foot of the current peak of the deeper traps. Then strong evidence is obtained for the emptying of holes from the deep traps, the Hall effect finally becoming n-type at high temperature as the dark conductivity dominates.

## 8.7 A mystery story: 'photochemical changes'

Occasionally special cases arise that have such a rich variety of characteristics that they are able to provide illustrations for a number of different phenomena, showing how they can be related. One such case is the discovery of particular traps in certain crystals of CdSSe and electron-irradiated CdS that fall into the general category sometimes known as 'photochemical changes.'[11] A condensed version of the analysis of the problem is presented here.

The effect is illustrated in Figure 8.12, showing the TSC curves for two crystals measured under two different conditions of photoexcitation: (1) excitation during the cooling cycle from high to low temperatures, and (2) excitation at low temperature only.[8,12] Attention focuses on the high temperature peak near room temperature that does not appear in the TSC curve unless photoexcitation occurred during cooling at temperatures above 180 K. The questions to be answered are: Why is photoexcitation needed while cooling? What kind of a trap is this? Can we analyze its properties using the techniques described above? Subsequent data concern the crystal of Figure 8.12($b$).

One assumption that can be made is that the trap distribution is a quasi-continuous trap distribution throughout, exhibiting strong re-trapping so that Fermi level analysis of the trap depths is justified. Measurements of photoconductivity decay were made at 18 different temperatures between 90 and 340 K. The results of the analysis of these decay curves are shown in Figure 8.13; excellent agreement is found between trap densities derived from decay data at different temperatures. A well-defined maximum is found for a trap depth of 0.49 eV. Similarly the TSC curves can be subject to Fermi level analysis and again the density of traps per $kT$ can be plotted as a function of trap depth; the results are essentially the same as those shown in Figure 8.13.

Rather surprising results are obtained, however, if the decay and TSC of the 289 K trap are considered separately in terms of a no-retrapping situation. If the measured decay curves are replotted on a semilog plot, it becomes clear that over the temperature range in which the 289 K trap dominates, i.e., between 244 and 304 K, each

Figure 8.12. TSC curves for (*a*) CdS crystal irradiated with 240 keV electrons, and then annealed at 200 °C under photoexcitation, and (*b*) a CdS$_{0.75}$CdSe$_{0.25}$ crystal. The solid curves indicate TSC measurements after photoexcitation while cooling from high temperatures; the dashed curves indicate TSC measurements after photoexcitation at low temperature only.[12]

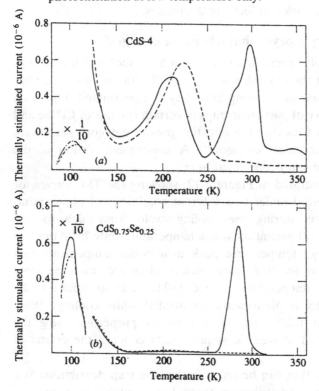

Figure 8.13. The trap density per unit volume per $kT$ as a function of trap depth as determined by a Fermi level analysis of the decay curves measured at 18 different temperatures, under the assumption that there is a quasi-continuous trap distribution subjected to strong retrapping emptying.[9]

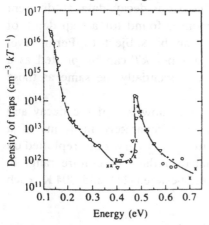

decay curve has a strong exponential component, and if the log of the exponential decay slope of these curves is plotted as a function of reciprocal decay temperature, as in Figure 8.14, a straight line is obtained over almost three orders of magnitude of decay slope. The corresponding trap has a depth of $E_c - E_t = 0.73\,\text{eV}$ and a relatively large capture cross section for electrons of $S_t = 1.3 \times 10^{-14}\,\text{cm}^2$. If this 289 K TSC peak really corresponded to such a trap without retrapping, then the TSC variation predicted by Eq. (8.58) should be observed. Figure 8.15 shows the theoretical TSC curve plotted according to Eq. (8.58) with $E_c - E_t = 0.73\,\text{eV}$, and $S_t = 1.3 \times 10^{-14}\,\text{cm}^2$, together with typical points taken from the measured TSC curve of Figure 8.12($b$). The agreement is so good that there can be no doubt but that the 289 K trap is a discrete trap that empties according to no-retrapping kinetics.

It can readily be shown that if the decay data for discrete traps with depth of 0.73 eV emptying without retrapping is analyzed as if they were a quasi-continuous distribution in terms of a Fermi level approach, exactly the results indicated in Figure 8.13 should be

Figure 8.14. Temperature dependence of the slopes of the exponential decay curves obtained by plotting the decay curves obtained between 244 and 304 K as $\ln n$ vs $1/T$. The data are consistent with no-retrapping decay of a discrete trap with depth of 0.73 eV and an electron capture cross section of $1.3 \times 10^{-14}\,\text{cm}^2$.[8]

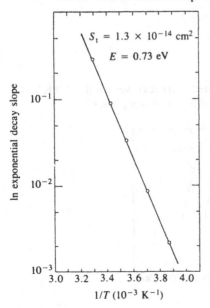

expected, i.e., a trap depth of 0.49 eV would be indicated instead of a trap depth of 0.73 eV. It is evident that serious errors can occur by using the wrong formulation for the trap emptying kinetics.

A plot of $\ln(T_m^2/\gamma)$ vs $1/T$ is independent of trapping kinetics, and should therefore provide an independent measurement of trap depth for the 289 K trap. Figure 8.16 shows the TSC curves for the 289 K peak for four different heating rates differing by a factor of 11. A plot of $\ln(T_m^2/\gamma)$ vs $1/T$ does give a straight line consistent with a trap

Figure 8.15. Comparison between the theoretical TSC curve calculated according to Eq. (8.58) for a discrete trap with depth of 0.73 eV and electron capture cross section of $1.3 \times 10^{-14}$ cm$^2$ emptying without retrapping, with the experimental TSC points for the 289 K TSC peak of Figure 8.12(*b*).[8]

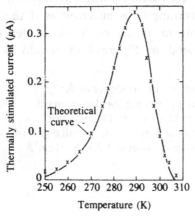

Figure 8.16. Thermally stimulated current curves for the 289 K TSC peak of Figure 8.12(*b*) for four different heating rates.[8]

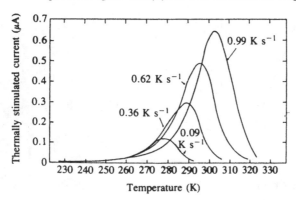

depth of 0.68 eV, which can be considered within experimental error of the 0.73 eV trap depth previously determined.

Although some questions have therefore been cleared up, some interesting puzzles remain. How can a trap with a large electron cross section be described by no-retrapping kinetics? An additional critical clue comes from the recognition that the measured TSC curves depend not only on the temperature of photoexcitation, but also on the wavelength of photoexcitation. Figure 8.17 shows the magnitude of the high-temperature TSC peak as a function of photoexcitation wavelength at 216 K, and compares the excitation spectrum obtained in this way with the excitation spectrum for photoconductivity. Major conclusions from measurements of this type are that (1) there is an optimum wavelength for filling these traps, (2) this optimum wavelength is not the wavelength corresponding to maximum photoconductivity, but corresponds to a smaller photon energy, and (3) this optimum wavelength is characteristic of a particular trap.

A plausible description of this phenomenon involving the 289 K

Figure 8.17. Excitation spectra for filling the high-temperature TSC peak (excitation at 216 K for 10 min) and for photoconductivity. The lower solid curve indicates the results of a measurement made for equal free electron density at each wavelength.[12]

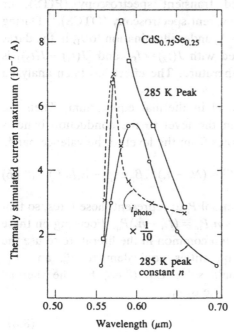

peaks in CdSSe crystals can now be summarized. Photoexcitation with
a certain wavelength 'dissociates' a neutral imperfection center into a
pair of shallow metastable electron and hole traps both occupied, the
most effective wavelength being about 590 nm at 216 K. This dis-
sociated center forms a stable trap-pair, either by lattice relaxation or
diffusion (temperature-dependent processes), producing traps with
much larger depths. Thermal emptying of an electron from the
electron trap simultaneously releases the hole from the hole trap,
resulting in collapse of the center back to its original configuration.
Once the electron is excited into the conduction band, the trapping
levels no longer exist and retrapping cannot occur. Photoexcitation
above the threshold temperature by wavelengths beyond the effective
excitation band does not bring about dissociation of the original
aggregate. Photoexcitation even with the optimum wavelength for the
dissociation of the center does not result in permanent stabilization of
the separated member if the temperature is below the threshold
temperature for lattice relaxation or diffusion.

## 8.8     Photoinduced transient spectroscopy

The rate window approach described in Section 8.1 has been
developed into a transient technique useful for high resistivity materials, known as photoinduced transient spectroscopy (PITS), or
sometimes as optical transient current spectroscopy (OTCS).[13] During
the decay of photocurrent from $J_L$ under illumination to $J_D$ in the dark,
two times $t_1$ and $t_2$ are selected with $J(t_2) \approx J_D$, and $[J(t_1) - J(t_2)]$ is
measured as a function of temperature. The effect has been analyzed
as follows.

Consider an imperfection level in thermal equilibrium. Then the
emission rate of electrons from the level to the conduction band is
equal to the emission rate of holes from the level to the valence band.

$$n_t N_c \beta_n \exp[-(E_c - E_t)/kT] = (N_t - n_t)N_v \beta_p \exp(-E_t/kT) \quad (8.66)$$

We have previously used the symbol $P_t$ to represent these rates, so that
Eq. (8.66) could be rewritten as $n_t P_{tn} = (N_t - n_t)P_{tp}$. Focusing on these
rates as 'emission' rates, it is often common in the literature to use the
symbols $e_n$ and $e_p$ for the quantities equivalent to $P_{tn}$ and $P_{tp}$.
Following this custom, we could solve Eq. (8.66) for the thermal
equilibrium (long decay time) value of $n_t$:

$$n_t(\infty) = N_t/(1 + e_n/e_p) \quad (8.67)$$

Under illumination

$$n_t e_n - \Delta n \beta_n (N_t - n_t) = (N_t - n_t) e_p - \Delta p n_t \beta_p \qquad (8.68)$$

giving for the initial value of $n_t(0)$ under illumination:

$$n_t(0) = N_t \{1/[1 + (e_n + \Delta p \beta_p)/(e_p + \Delta n \beta_n)]\} \qquad (8.69)$$

The current generated by the trap emptying is

$$J(t) = C\{e_n n_t(t) + e_p [N_t - n_t(t)]\} \qquad (8.70)$$

where $C$ is a geometrical parameter involving the penetration depth of the light. The transient current is

$$\partial J(t) = [J(0) - J(\infty)] \exp(-t/\tau) \qquad (8.71)$$

where

$$J(0) = C\{e_n n_t(0) + e_p[N_t - n_t(0)]\} \qquad (8.72a)$$

$$J(\infty) = C\{e_n n_t(\infty) + e_p[N_t - n_t(\infty)]\} \qquad (8.72b)$$

so that

$$\partial J(t) = C(e_n - e_p)[n_t(0) - n_t(\infty)] \exp(-t/\tau) \qquad (8.73)$$

with

$$\tau = 1/(e_n + e_p) \qquad (8.74)$$

In general, therefore, incorporating Eqs. (8.67) and (8.69),

$$\partial J(t) = C(e_n - e_p)(N_t\{1/[1 + (e_n + \Delta p \beta_p)/(e_p + \Delta n \beta_n)]\} \\ - N_t/(1 + e_n/e_p)) \exp(-t/\tau) \qquad (8.75)$$

For high excitation intensities, $e_n \ll \Delta p \beta_p$, $e_p \ll \Delta n \beta_n$, and $\Delta n \approx \Delta p$. Then

$$\partial J(t) = C(e_n - e_p)N_t\{[1/(1 + \beta_p/\beta_n)] - 1/(1 + e_n/e_p)\} \exp(-t/\tau) \qquad (8.76)$$

For an electron trap, $\beta_n \gg \beta_p$, $e_n \gg e_p$, and

$$\partial J(t) = C e_n N_t \exp(-e_n t) \qquad (8.77)$$

This is the typical result previously obtained for trap decay without retrapping, with the trap filled at $t = 0$ and empty at $t = \infty$. Calculation of $d[\partial J(t)]/dT = 0$ gives $t$ for $T_m$: $1/t = e_n$. Thus for the time interval $t_1$

$$1/t_1 = N_c S_n v_n \exp[-(E_c - E_t)/kT_m] \qquad (8.78)$$

Rate window differences are measured as a function of temperature for different values of the sampling time $t_1$, and then a plot of

$\ln(T^2/e_n)$ vs $1/T$ gives the trap depth $(E_c - E_t)$, taking into account the $T^{3/2}$ dependence of $N_c$ and the $T^{1/2}$ dependence of $v_n$. A typical plot is shown in Figure 8.18 for a particular trap in a GaAs vapor phase epitaxy layer.[13]

In the event that $e_p \gg e_n$ for hole traps, a similar analysis holds. If $e_p \approx e_n$, the situation is more complicated; this case usually corresponds to levels near the middle of the gap, and since $1/\tau = (e_n + e_p)$, $(E_c - E_t)$, cannot be determined. The method does not yield values for the trap density $N_t$ in a simple way.

## 8.9    Deep level transient spectroscopy

Deep level transient spectroscopy (DLTS) is strictly speaking not a photoelectronic analysis technique since no photoexcitation is usually involved. Also to be able to detect both majority and minority carrier traps, it must be possible to prepare effective p⁺–n or n⁺–p junctions. Since it and its many variations are, however, the most widely used methods for characterizing imperfections in more conducting semiconductors today, some description of the method should be included here. It involves the measurement of the junction capacitance transient, due to levels filled or emptied by injection as

Figure 8.18. PITS curves measured for a particular trap in vapor phase epitaxial GaAs, showing the influence of a change in the delay time $t_1$. The inset is a plot of $\ln(T^2/e_n)$ vs $1/T$, compared with the dashed line for trap EL6 measured independently by deep level transient spectroscopy.[5]

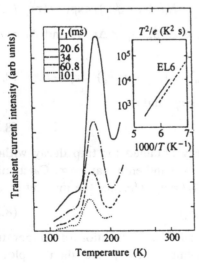

they return to equilibrium, as a function of temperature.[14,15] In actual operation, this capacitance transient is recorded by suitable cyclical repetition of the transient and the data are subject to multichannel signal-averaging to provide high-quality data.

The capacitance of a junction is given by Eq. (1.26) and was discussed in Section 7.4.

$$(C/A)^{-2} = 2(\phi_D - \phi_a)/(\epsilon_r \epsilon_0 q N_S) \tag{8.79}$$

where $N_S$ is the total space charge density.

An example of the application of DLTS to a donor level can be summarized as follows. Define a depletion layer width in the presence of a steady reverse bias $\phi_R$. Decrease the bias suddenly to 0 V so that the depletion layer narrows, the junction capacitance increases, and majority carriers are trapped. Return to a reverse bias $\phi_R$ so that the capacitance is decreased below its initial value by an amount $\Delta C_0$ (since now the density of charged donors has been reduced by the trapped majority carriers) and observe the decay $\Delta C(t) = \Delta C_0 \exp(-e_n t)$ by thermal emptying of levels filled by the majority carrier pulse. Repeat this measurement as a function of temperature. Figure 8.19 shows a schematic of the process obtainable with a $p^+$–n junction. Similar effects associated with injection of minority carriers with a $p^+$–n junction are shown in Figure 8.20.

A forward bias voltage pulse injects minority carriers in $p^+$–n or $n^+$–p junctions; such injection of minority carriers is not possible in a Schottky barrier or MOS junction. The sign of $\Delta C$ depends on the carrier capture and the conductivity-type of the material. The sign is negative when majority carrier electrons fill electron traps in n-type material, or holes fill hole traps in p-type material. The sign is positive when minority carrier holes fill traps in n-type material, or electrons fill traps in p-type material.

The application of the rate window approach to this measurement is illustrated in Figure 8.21. The capacitance difference is given by

$$[C(t_1) - C(t_2)] = \Delta C[\exp(-e_n t_1) - \exp(-e_n t_2)] \tag{8.80}$$

To find the temperature at which the capacitance difference is a maximum, differentiate with respect to $T$ and set equal to zero to obtain for the maximum:

$$\tau = 1/e_n = (t_1 - t_2)/\ln(t_1/t_2) \tag{8.81}$$

$t_1$ and $t_2$ can be varied to give different values for $T_m$ as in PITS in Section 8.7.

Figure 8.19. Capacitance transient due to a majority carrier trap in a
$p^+$–n junction. The insets show the charge state of the defect level
and the width of the space charge region (unshaded portion) at
successive times before and during the transient.[15]

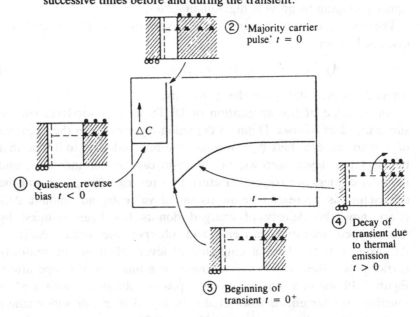

Figure 8.20. Capacitance transient due to a minority carrier trap in a
$p^+$–n junction. The insets show the charge state of the defect level
and the width of the space charge region (unshaded portion) at
successive times before and during the transient.[15]

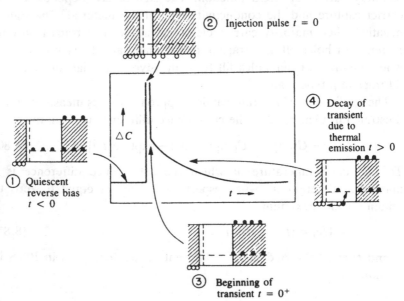

Illustrations of typical DLTS spectra are given in Figures 8.22 and 8.23. Figure 8.22 shows typical electron and hole traps in electron irradiated Si doped with B.[16] Figure 8.23 shows how several specific imperfection levels in n-$Al_xGa_{1-x}As$ vary with $x$ values of 0, 0.15 and 0.28.[17] As the bandgap increases with increasing $x$, the imperfection levels associated with Cu, Fe, and the commonly occurring A and B defects in GaAs grown by liquid phase epitaxy, all shift to higher energies and hence to DLTS maxima at higher temperatures.

Figure 8.21. The rate window approach to measurement of DLTS. The output of a double-boxcar integrator corresponds to the average difference of the amplitudes at the sampling times $t_1$ and $t_2$.[15]

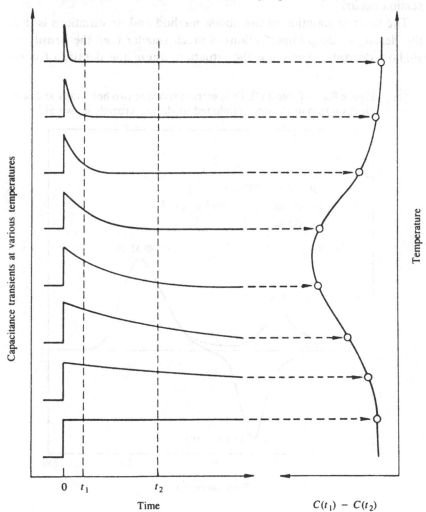

Capacitance transients at various temperatures

Temperature

0   $t_1$          $t_2$

Time

$C(t_1) - C(t_2)$

In spite of its great power, there are a number of problems that arise with the use of DLTS in materials that depart from near-ideal semiconductor properties. If the density of imperfections is very large, the decay becomes non-exponential and the simple interpretation is no longer adequate. If the doping is non-uniform with distance in the material, problems of interpretation arise. Failure to include actual temperature dependences of the capture cross section may lead to errors in the determination of the level energy. If measurements are carried out in regions with high electric fields, there is always the possibility that the magnitudes of the emission rates are affected by these fields. Problems arise when the signal from two or more overlapping deep levels with different trap depth and capture cross section occur.

The basic assumption in the above method and its variations is that the density of deep imperfections is much smaller than the density of shallow dopants. Analyzing the situation where the density of deep

Figure 8.22. Typical DLTS spectrum showing two hole traps and one electron trap in electron-irradiated, B-doped, crucible grown Si.[16]

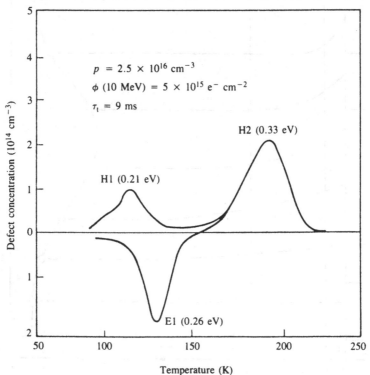

Figure 8.23. DLTS hole emission spectra for three samples of n-Al$_x$Ga$_{1-x}$As for $x = 0$, 0.15 and 0.28.[17]

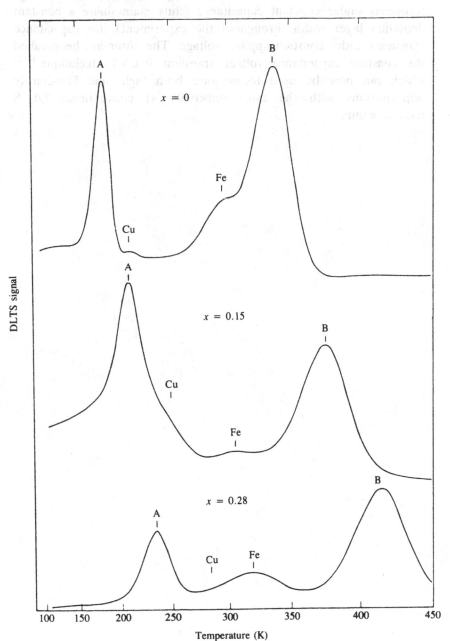

imperfections is comparable to or even greater than the dopant density requires a basic modification in the technique: substitution of voltage transients under constant capacitance (thus maintaining a constant depletion layer width throughout the experiment), for capacitance transients under constant applied voltage. The latter has been called the constant capacitance voltage transient (CCVT) technique,[18-21] which can now be used to measure both high- and low-density imperfections with the same sensitivity as conventional DLTS measurements.

# 9

# Photoeffects at grain boundaries

When semiconductors are prepared in thin film form, single crystal films may be obtained provided that it is possible to achieve epitaxial growth of the film on a single crystal substrate. But if semiconductor films are deposited under non-epitaxial conditions or on amorphous substrates, then it is most likely that they will occur in polycrystalline form.

As introduced in Section 2.7, the boundary between two crystalline grains in a polycrystalline film is describable by a potential barrier such as is shown in Figure 2.10. The grain boundary barrier is described in terms of a diffusion potential $-q\phi_b$, the energy difference between the top of the barrier and the semiconductor conduction band edge, and a grain barrier height $-q\phi_{gb}$, the energy difference between the top of the barrier and the Fermi level in the semiconductor. It follows that in an n-type semiconductor such as is pictured in Figure 2.10 $q\phi_{gb} = q\phi_b + (E_c - E_F)$.

The basic results of Section 2.7 can be summarized here. For $q\Delta\phi/kT \ll 1$, the total current is given by

$$J = (nq^2\Delta\phi\, v/kT)\exp(-q\phi_b/kT) \qquad (2.59)$$

The conductivity is given by

$$\sigma = (nq^2 d_g v/kT)\exp(-q\phi_b/kT) \qquad (2.60)$$

If $\sigma = nq\mu^*$,

$$\mu^* = (qv d_g/kT)\exp(-q\phi_b/kT) \qquad (2.61)$$

Under illumination

$$\Delta\sigma = q\mu^*\Delta n + nq\Delta\mu^* \qquad (2.62)$$

with

$$\Delta\mu^* = -[q^2 v d_g/(kT)^2]\Delta\phi_b \exp(-q\phi_b/kT) \tag{2.63}$$

The grain boundary interface states (see Figure 2.10) below the Fermi level are negatively charged, and the grain boundary interface is surrounded on both sides by semiconductor depletion regions with a positive charge. If $N_{gb}^-$ is the density of charged interface states, $N_D^+$ the density of charged donors in the semiconductor, and $w_d$ the depletion layer width,

$$N_{gb}^- = 2N_D^+ w_d \tag{9.1}$$

Now the diffusion potential from Eq. (1.24) is given by

$$\phi_b = (qN_D^+ 2\epsilon_r\epsilon_0)w_d^2 \tag{9.2}$$

From Eq. (9.1) $N_D^+ w_d^2 = (N_{gb}^-)^2/4N_D^+$, so that

$$\phi_b = q(N_{gb}^-)^2/(8\epsilon_r\epsilon_0 N_D^+) = \phi_{gb} - (E_c - E_F)/q \tag{9.3}$$

The result will be different in the special case that the grains are fully depleted. Then $w_d \to d_g/2$ and

$$\phi_b = (qN_D^+ d_g^2)/(8\epsilon_r\epsilon_0) \tag{9.4}$$

The value of $\phi_{gb}$ is determined from $N_{gb}^- = N_D^+ d_g$.

In this chapter we discuss examples of effects observed with polycrystalline thin films to illustrate the kind of phenomena encountered.

## 9.1    Small grain vs large grain effects: CdTe films

The effect of grain boundaries, and hence of grain size, on the electrical properties of thin semiconductor films can be demonstrated by comparing the properties of small-grain (1 $\mu$m) CdTe films deposited by hot wall vacuum evaporation (HWVE) on glass substrates with essentially single crystal, epitaxial films deposited on (111) single crystal BaF$_2$ substrates.[1]

Figure 9.1 shows the measured variation of the resistivity in dark and light for CdTe films deposited by HWVE as a function of the source temperature of co-evaporated indium dopant. The film resistivity decreases with increasing In source temperature until some temperature is reached at which the resistivity appears to saturate. The resistivities for the epitaxial films (corresponding to a hole density of $10^{16}$–$10^{17}$ cm$^{-3}$) are orders of magnitude smaller than those of the small-grain polycrystalline films (corresponding to hole densities of about $10^{11}$ cm$^{-3}$). The major differences between the two types of film can be understood by realizing that the small-grain polycrystalline films exhibit the properties of totally depleted grains, whereas the epitaxial films do not.

Further consequences of this difference in structure can be seen by making photo-Hall measurements as a function of photoexcitation intensity. Figure 9.2 shows the results of such measurements on both types of film as a function of the film resistivity in the light. The small-grain polycrystalline films show a change in electron density by

Figure 9.1. Dark and light resistivity of CdTe:In films grown on glass substrates and on single crystal BaF$_2$ substrates as a function of the In source temperature.[1] Data points correspond to films on glass in the dark (filled circles) and in the light (open circles); films on BaF$_2$ grown with a substrate temperature of 480 °C in the dark (filled triangles) and in the light (open triangles); films on BaF$_2$ grown with a substrate temperature of 450 °C in the dark (filled squares) and in the light (open squares).

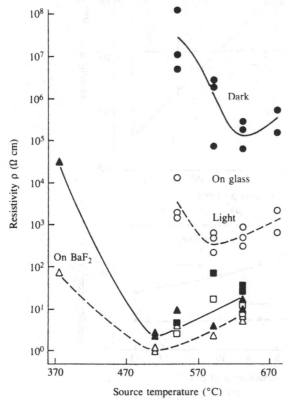

orders of magnitude, but no change in the mobility. The epitaxial films show an increase in the electron mobility under illumination but no change in the electron density. If it is assumed that the major effect of illumination is to decrease the density of grain boundary states $N_{gb}$,[2,3] then a grain boundary model[4,5] that predicts the variation of electron density and mobility with $N_{gb}$ can be used to indicate approximately the expected variation with illumination, with the results given in Figure 9.3. For totally-depleted small-grain films, the increase in

Figure 9.2. Variation of the electron density and mobility under optical excitation as a function of the resistivity in the light for CdTe films deposited on (*a*) glass substrates, and (*b*) BaF$_2$ substrates.[1]

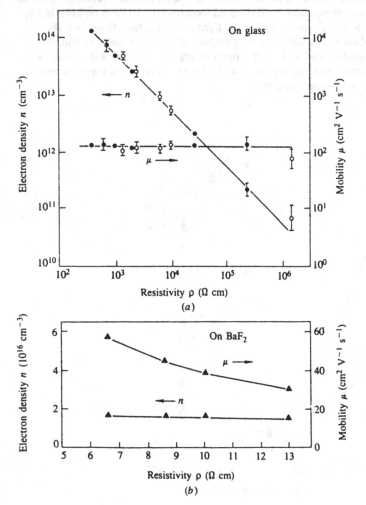

photoconductivity with illumination results almost completely from an increase in the electron density, whereas for epitaxial films, the increase in photoconductivity with illumination results almost completely from an increase in the electron mobility, consistent with the results of Figure 9.2.

## 9.2      Direct measurements of grain boundary properties: CdTe films

The properties of grain boundaries in CdTe have been examined further in more detail by carrying out experiments on single grain boundaries in bicrystals.[6] The resistance associated with such grain boundaries is typically 3–5 orders of magnitude larger than would be found for the bulk CdTe material. The exponential dependence of grain boundary resistivity on reciprocal temperature is illustrated in Figure 9.4 and indicates that carrier transport is by thermionic emission over (or thermally assisted tunneling through) a potential barrier present at the boundary. The height of the barrier decreases monotonically with increasing light intensity, with a cor-

Figure 9.3. Calculated values of the electron density and mobility as a function of the total density of grain boundary states $N_{gb}$. Solid lines are for films on glass, dashed lines for films on BaF$_2$.[1]

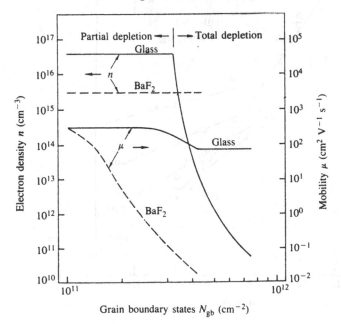

Grain boundary states $N_{gb}$ (cm$^{-2}$)

responding increase in carrier mobility as charged boundary states are filled with photoexcited minority carriers.

The grain boundaries in these bicrystals and in other polycrystalline films similarly tested can be temporarily passivated. The effects of grain boundaries in p-type CdTe can be reduced by heat treatment in molecular $H_2$, a $H_2$ plasma, or by Li diffusion. Figure 9.5 shows the temperature dependence of the current crossing a p-CdTe bicrystal in dark and light, before and after passivation by heating in $H_2$. A marked decrease in activation energy with corresponding increase in current magnitude is found upon passivation, but such passivation decays with a time constant of the order of a few weeks.

The actual energy distribution of the grain boundary states was

Figure 9.4. Grain boundary conductivity as a function of temperature in the dark and for several different illumination intensities ($I_0 =$ 60 mW cm$^{-2}$ white light) for the grain boundary in a p-type CdTe bicrystal.[6] The inset shows the variation of the activation energy with light intensity at low temperatures.

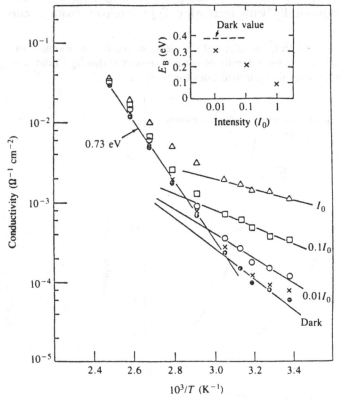

measured before and after passivation, by measurements of grain boundary current vs voltage measurements,[7] and of grain boundary photocapacitance and/or photoconductivity.[8] Results for a p-type bicrystal of CdTe are shown in Figure 9.6. For all samples measured, there was a monotonic decrease in state density with energy separation from the band edge, similar to what is observed for Si bicrystals and amorphous materials. Figure 9.6 shows that the effects of passivation are to decrease the density of these grain boundary states, as expected.

Although detailed investigation of individual grain boundaries cannot be carried out in polycrystalline p-type CdTe films, the effects of passivation can be tested. Results similar to those shown in Figure 9.5 were found on heat treatment in $H_2$ (but not for heat treatment in

Figure 9.5. Current across a grain boundary in a p-type CdTe bicrystal as a function of temperature, in dark and light, before and after passivation by heating in $H_2$.[6]

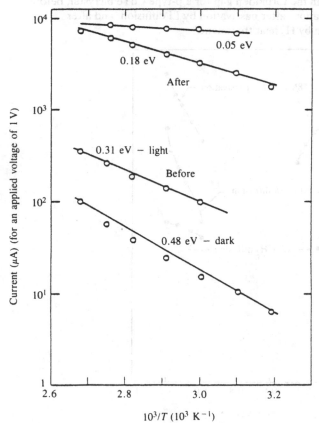

either vacuum or air) or Li diffusion in the polycrystalline films. When n-type CdTe bicrystals or polycrystalline films are examined, similar effects can be observed, except that for the n-type films it is heat treatment in $O_2$ not $H_2$, that produces temporary passivation of the grain boundary states.

## 9.3    CdS films

Thermoelectric power and photothermoelectric power measurements have been applied to n-type CdS films. Figure 9.7 shows the temperature dependence of electron density and mobility of as-prepared films deposited by spray pyrolysis;[9] it has many similarities to the data of Figure 9.4 for a p-CdTe bicrystal. Electron densities both in the dark and under various photoexcitation intensities are independent of temperature; the dark electron mobility is thermally

Figure 9.6. Density of grain boundary interface states as a function of energy in the forbidden gap for a p-type CdTe bicrystal, before any passivation, after passivation by Li diffusion, and after passivation by $H_2$ heat treatment.[6]

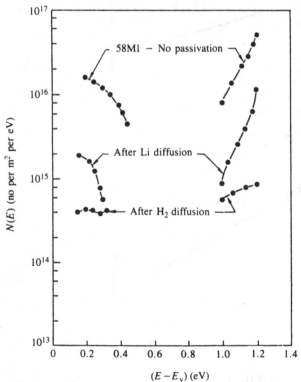

activated, but mobility under photoexcitation is independent of temperature at low temperatures, depending mainly on the excitation intensity. Photoconductivity is caused mostly by an increase in the electron mobility.

### Heat treatment in $H_2$

When such films are heat treated in $H_2$ major increases in conductivity can be produced corresponding to both an increase in electron density and an increase in mobility.[10] Figure 9.8 shows the values of these quantities before and after a 5 min heat treatment at 400 °C. The conductivity becomes degenerate, the electron mobility is increased by an order of magnitude to a value approaching that found

Figure 9.7. Electron density and mobility as a function of temperature for a solution-sprayed CdS film in the dark and under photoexcitation ($L = 1.0$ corresponds to white light excitation of $32 \, mW \, cm^{-2}$).[9]

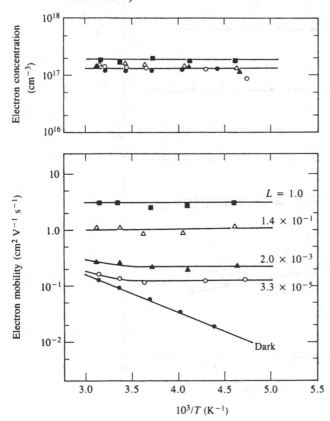

in single crystals, and the mobility activation energy is significantly reduced from 0.083 eV to 0.015 eV. Neither before nor after heat treatment is there any effect of illumination on the electron density in these films. A major effect of such a $H_2$ heat treatment is to increase the non-stoichiometry of the films, producing excess Cd donors, increasing the free electron density in the grains, and reducing the grain boundary barrier heights. X-ray diffraction measurements also indicate an increase in grain size as a result of the heat treatment.

Figure 9.8. Temperature dependence of the electron density and the electron mobility of CdS films deposited by spray pyrolysis before and after heat treatment in $H_2$ for 5 min at 400 °C. Before heat treatment (filled and open circles), and after heat treatment (filled and open squares). Solid curves are in the dark; dashed curves are under illumination.[10]

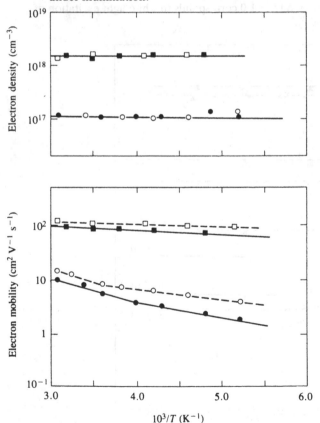

### O₂ adsorption effects

n-type films are also especially susceptible to $O_2$ adsorption effects.[11-13] If there had been any such effects on the sample of Figure 9.8, the $H_2$ heat treatment would have removed them as well. Physically adsorbed $O_2$ can take an electron from the material to form a chemically adsorbed surface state, leaving behind a depletion region near the surface of the material. If the film is thin enough, this depletion region can extend throughout the entire width of the film. Figure 9.9 shows the change in conductivity of a CdS film due to heat treatment at 120 °C in air or in vacuum.[9]

Figure 9.9. Dark conductivity as a function of temperature for an as-prepared solution-sprayed CdS film after heat treatment at 100 °C in air for 18 h and after subsequent heat treatment at 120 °C in vacuum for 18 h.[9] Also shown is the light conductivity under different photoexcitation intensities after the vacuum treatment ($L = 1.0$ corresponds to white light excitation of 50 mW cm$^{-2}$).

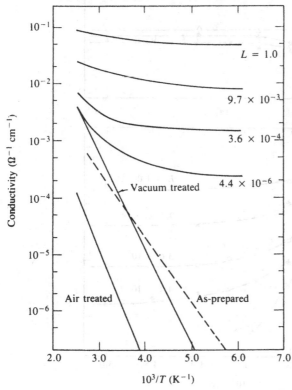

After the final vacuum heat treatment of Figure 9.9, the film was subjected to photothermoelectric analysis to determine values of electron density and mobility in the dark and under photoexcitation, as shown in Figure 9.10, which is similar to the as-prepared data given in Figure 9.7. The high-temperature portion of the mobility curves of Figure 9.10 can be interpreted to indicate a decrease in mobility activation energy from 0.46 eV in the dark to 0.22 eV for full photoexcitation intensity.

This decrease in activation energy may be fairly well fit by considering that the reduction in grain boundary barrier height caused

Figure 9.10. (a) Electron density and (b) electron mobility in dark and light for the CdS film of Figure 9.9 after the vacuum heat treatment.[9] The apparent decrease in electron density at high temperatures is an artifact of the thermoelectric power measurements.

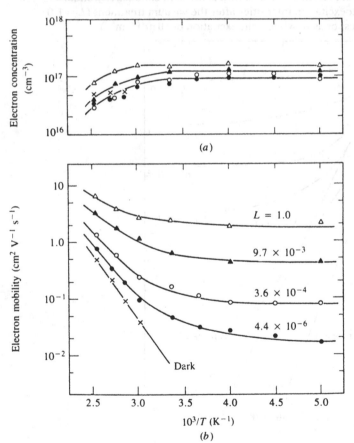

by light results in an increased probability for tunneling through the barrier. Holes generated by photoexcitation near the grain boundaries are trapped at the grain boundaries with an effectiveness depending on the ratio of hole lifetime and the transit time to the boundaries. The rate of hole generation depends on the photoexcitation intensity. Holes trapped at the grain boundaries decrease the barrier height and barrier width. The decrease in barrier height is indicated by the decrease of mobility activation energy with photoexcitation, and the decrease in the barrier width increases the tunneling probability of electrons passing through the barrier, a process that becomes predominant at low temperatures. Figure 9.7 shows that a tunneling process dominates the electrical transport at high-intensity excitation over the whole temperature range measured. At low excitation intensities, thermal excitation over the barriers and tunneling are parallel processes.

### Effect of dc electric field during heat treatment

A dramatic increase in conductivity is found to occur in these films if they are heated to 100 °C with a dc electric field applied. No effect on the conductivity is found if a 60 Hz ac field is used. Dynamic curves of dark conductivity as a function of time at 100 °C are shown in

Figure 9.11. Dark conductivity at 100 °C in vacuum of an as-prepared solution-sprayed CdS film as a function of time with dc voltages as indicated applied between electrodes.[9]

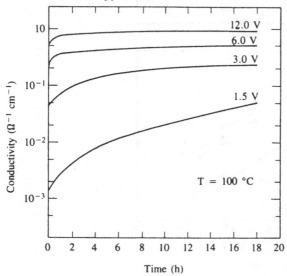

Figure 9.11 for different applied dc voltages. After each of the heat treatments with dc field shown in Figure 9.11, the temperature dependence of electron density and electron mobility in the dark was measured with the results shown in Figure 9.12. The results indicate that both electron density and mobility are increased by the application of a dc field during heat treatment.

This effect can be dramatically increased in magnitude if Cu acceptor impurities are diffused into as-deposited films. The effect of the Cu diffusion is to decrease the dark electron density by several orders of magnitude and to decrease the dark electron mobility by at least one order of magnitude. In the absence of dc electric fields, heat treatment at 100 °C in air or vacuum produces generally small changes

Figure 9.12. Dark electron (*a*) density and (*b*) mobility as a function of temperature of an as-prepared solution-sprayed CdS film after (1) vacuum treatment at 100 °C for 35 h without applied voltage, and after vacuum treatment at 100 °C for 18 h with applied dc voltages of (2) 1.5, (3) 3.0, (4) 6.0 and (5) 12.0 V.[9]

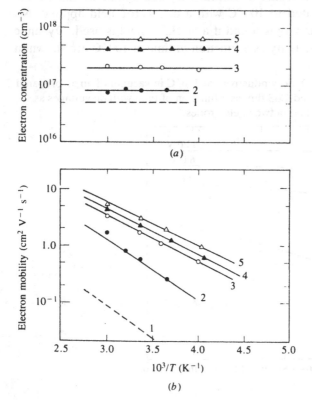

in the Cu-diffused films as compared to the as-prepared films. When heat treatment is carried out with an applied dc electric field, however, the effects are large as shown in Figure 9.13, which may be compared with Figure 9.11 for the as-deposited film. The dark conductivity and photoconductivity as a function of temperature after each dc field treatment are shown in Figure 9.14. The dark conductivity of the film increases by eight orders of magnitude after four consecutive dc field treatments. These effects associated with heat treatment in the presence of applied dc electric fields appear to be describable in terms of field-assisted diffusion of charged acceptor-like impurities in the film to the anode.

Figure 9.13. Dark conductivity at 100 °C of a Cu-diffused solution-sprayed CdS film as a function of time with indicated voltages applied between end electrodes.

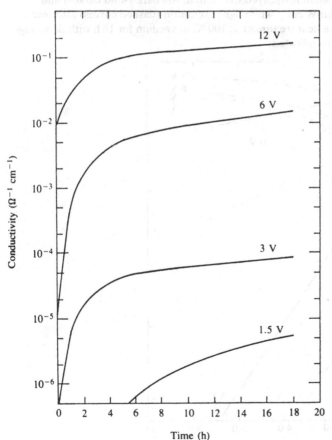

## 9.4    PbS films

PbS films have played an important role as infrared detectors for many years. The bandgap of PbS is given approximately by $E_G = 0.28 + 4 \times 10^{-4}T$ eV, which has a value of 0.40 eV at 300 K, corresponding to an absorption edge at 3.1 $\mu$m. In the early years of its development, considerable debate existed as to whether the photoconductivity was due primarily to a change in carrier density with photoexcitation, or to a change in carrier mobility. One of the interesting results of research is that, although the PbS films show characteristic polycrystalline film behavior in many ways, photoconductivity in the PbS films is controlled by a sensitizing center effect like that described in Chapter 5.[14,15]

Figure 9.14. Conductivity as a function of temperature of a Cu-diffused solution-sprayed CdS film in the dark (solid curves) and under 50 mW cm$^{-2}$ white light excitation (dashed curves) after each successive heat treatment at 100 °C in vacuum for 18 h with dc voltage applied as indicated.[9]

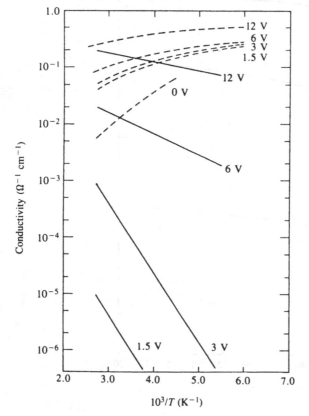

These films are traditionally prepared by precipitation from a suitable solution (Pb(CH$_3$CO$_2$)$_2$, thiourea, NaOH, and an oxidant) onto a substrate; the addition of the oxidant is crucial to producing a high photosensitivity. Films prepared without added oxidant show strong O$_2$ adsorption effects, but only a very small response to photoexcitation (about $10^{-3}$ that in material with oxidant added). We will discuss here only those 'standard' films prepared with oxidant.

### Effects of O$_2$ adsorption and desorption in the dark

Like CdS films, PbS films also show effects due to adsorption and desorption of O$_2$; since the PbS films are *p*-type, however, adsorption of O$_2$ produces an increase in hole density and film conductivity. Figure 9.15 shows the temperature dependence of hole conductivity (derived from the Hall coefficient $R_H$) for PbS films with absorbed O$_2$ and after vacuum heat treatment at two different temperatures (colloquially referred to as a vacuum 'fix' on the figure). The corresponding temperature dependence of Hall mobility is shown in Figure 9.16. Desorption of O$_2$ by vacuum heat treatment increases the resistivity, decreases the free-hole density, and increases the high-temperature activation energy for the hole density. The hole mobility is significantly reduced by desorption of O$_2$, but this reduction is primarily in the low-temperature range. The measured Hall mobility is very small compared to the value of $500 \, \text{cm}^2 \, \text{V}^{-1} \, \text{s}^{-1}$ expected at room temperature for single crystal PbS. In addition, the Hall mobility is thermally activated with an activation energy of 0.08 eV in the high temperature range and a smaller value of 0.04 eV at low temperatures.

As a test of the transport measurements, both Hall effect and thermoelectric power measurements were made as a function of temperature for a relatively high conductivity state, as described in Section 7.3. An analysis of an inhomogeneous model for PbS layers shows that the hole density determined by the Hall effect could be considerably less than the hole density in the grains of the layer, but that the carrier density determined by the thermoelectric effect is equal, under almost all circumstances to that in the grains.[16] It was found that a good correlation between the two kinds of data could be obtained over a reasonable temperature range if the value of $K$ in the Hall measurement were 1.93, and the value of $A$ in the thermoelectric power measurement were 4; these values correspond in an ideal case to scattering by charged imperfections. This correlation indicates that the hole density determined by Hall and thermoelectric power measurements is indeed the hole density in the grains.

Figure 9.15. Temperature dependence of the hole density (measured as $1/R_He$) for a PbS film in the dark, as initially received, and after the indicated vacuum heat treatments ('fixes').[14] Data reported by Petritz *et al.*[17] are also included. The encapsulated sample is one produced by placing a sapphire plate on top of the layer with a thermal-setting resin between.

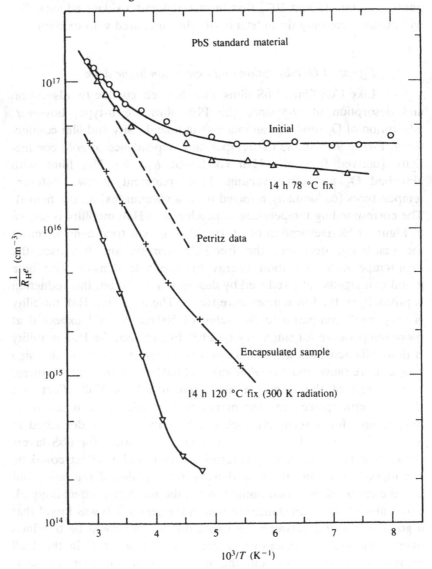

*Photoconductivity*

Figure 9.17 shows the results of photo-Hall measurements made as a function of temperature as initially received (with adsorbed $O_2$) and after a severe vacuum heat treatment, in the dark and at different photoexcitation intensities. The corresponding Hall mobilities are given in Figure 9.18. The proportional change in hole density with photoexcitation is greatly increased by vacuum heat treatment; the proportional change in Hall mobility is also increased by vacuum heat treatment but much less dramatically. It is only at low temperatures in the initial high-conductivity state that the proportional change in Hall

Figure 9.16. Temperature dependence of Hall mobility, corresponding to the hole density data given in Figure 9.15.[14]

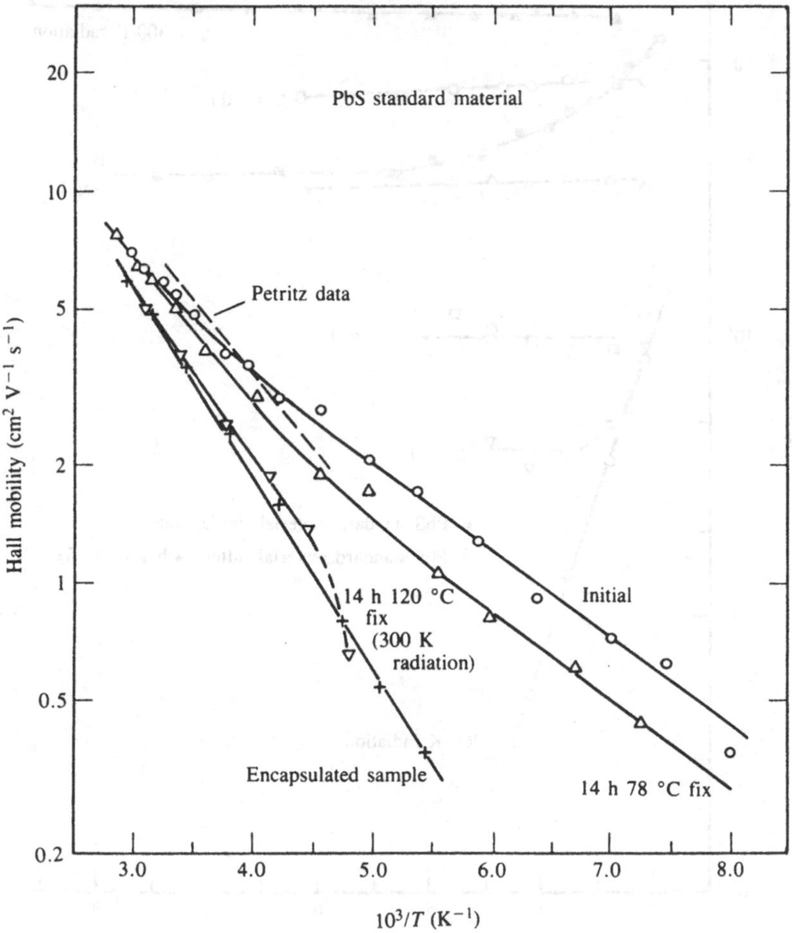

mobility is comparable to the proportional change in hole density. Thus photoconductivity in these oxidant-treated PbS layers is primarily due to an increase in hole density with only a small contribution due to an increase in hole mobility. In fact the effect of photoexcitation on the hole mobility in these films is so small, compared to that in CdS

Figure 9.17. Temperature dependence of hole density (measured as $1/R_He$) for a PbS film, as initially received and after a severe vacuum heat treatment ('fix'), for various photoexcitation intensities. Intensities $L$ are in units of mW cm$^{-2}$.[14]

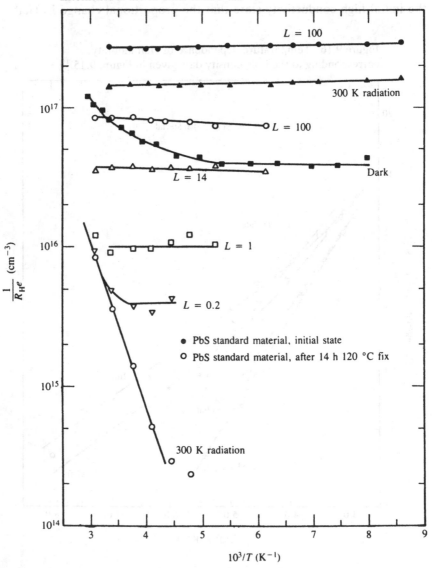

Figure 9.18. Temperature dependence of Hall mobility, corresponding to the hole density data given in Figure 9.17.[14]

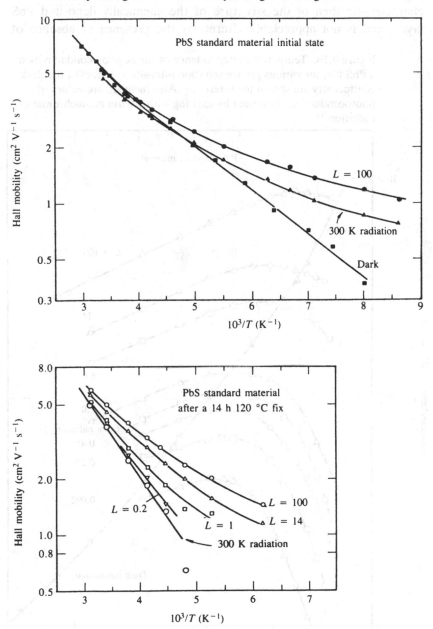

(Figure 9.10), for example, that it appears that the intergranular regions in the PbS are structurally controlled. The mobility behavior is characteristic then of the structure of the chemically deposited PbS layer, and is not appreciably altered by the presence or absence of

Figure 9.19. Temperature dependence of the ac photoconductivity of a PbS film for various photoexcitation intensities. Values of the dark conductivity are shown for reference. Also included are values of photoconductivity obtained by exciting with 2.6 $\mu$m monochromatic radiation.[14]

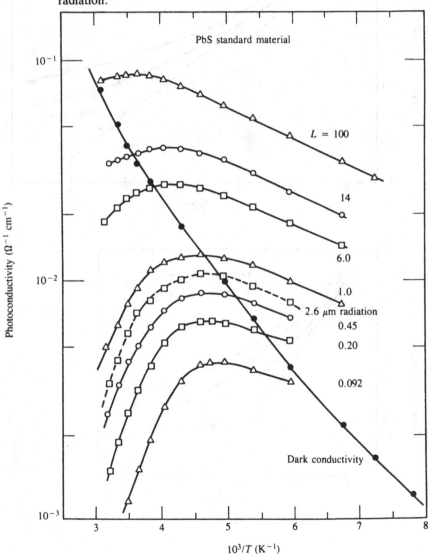

oxidant in the deposition, adsorption or deposition of $O_2$, or the thickness of the layer deposited.

The temperature dependence of the ac photoconductivity (14 Hz) is given for several different photoexcitation intensities in Figure 9.19. A maximum in the photoconductivity as a function of temperature is followed by thermal quenching of photoconductivity, corresponding to an activation energy for thermal quenching of carrier density of 0.22 eV.

As another check on the behavior, the photoconductivity decay time was measured for four different low photoexcitation intensities, as shown in Figure 9.20. The fact that the decay time becomes independent of excitation intensity in the thermal quenching region indicates that the temperature dependence is that of the hole lifetime. The activation for thermal quenching of the decay time is again 0.22 eV. It may be concluded, therefore, that photoconductivity in the 'standard' material is due to the effects of sensitizing centers associated with the oxidant, lying 0.22 eV below the conduction band.

Figure 9.21 presents a specific model for the band structure of the granular and intergranular portions of the PbS layer. It is proposed that barriers of height 0.08 eV are associated with grain boundaries. The transport of free holes past these barriers takes place predominantly by (i) thermal excitation over the barrier in the high-temperature region, or (ii) tunneling through the barrier in the low-temperature region. When an electron–hole pair is created by photoexcitation, the electron is trapped at the donor-like sensitizing center, and the hole is free to conduct. Photomodulation of the mobility is always an increase in the tunneling contribution to the total transport of holes, caused by decreasing the depletion layer width associated with the barrier through the trapping of negative charge at the barrier.

### Effects of $O_2$ adsorption: *time dependence*

In order to measure variations in the effects due to $O_2$ adsorption as a function of time, PbS layers were given a standard vacuum heat treatment in the dark to desorb $O_2$, and then were exposed to photoexcitation and air. The conductivity, Hall effect, and photoconductivity were measured as a function of time during photoexcitation. Measurements were also made for the sake of comparison of the variation of these quantities after a vacuum heat treatment upon exposure only to air without photoexcitation. In every case, photoexcitation is required for any measurable change.

Figure 9.22 shows the variation of measured hole density and Hall mobility as a function of time during photoadsorption of $O_2$ and as a function of time subsequently during desorption of $O_2$ in vacuum and dark. The increase in conductivity associated with photoadsorption of $O_2$ occurs through an increase in both hole density and hole mobility, although the time dependence of the two variations is not the same.

Figure 9.20. Temperature dependence of the photoconductivity decay time (time required for the photoconductivity to decrease to $1/e$ of its original value) for various low intensities of photoexcitation.[14]

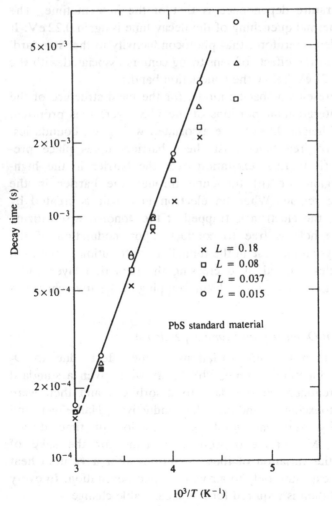

$10^3/T$ (K$^{-1}$)

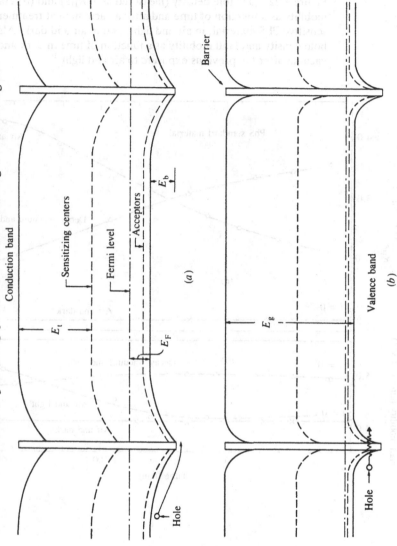

Figure 9.21. Band picture for one grain of photosensitive PbS material; (a) for large depletion widths associated with the intergrain barriers, and (b) for small depletion layer widths. $E_t = 0.22$ eV is the energy required to free an electron from a sensitizing center, $E_F$ the height of the Fermi level above the valence band, $E_g$ the bandgap, and $E_b = 0.08$ eV the intergrain barrier height.[14]

Figure 9.22. (*a*) Hole density (measured as $1/R_H e$) and (*b*) Hall mobility as a function of time and after a vacuum heat treatment for a sensitive PbS material, in air and light, and in air and dark. Also the hole density and Hall mobility as a function of time in dark and vacuum after the previous exposure to air and light.[14]

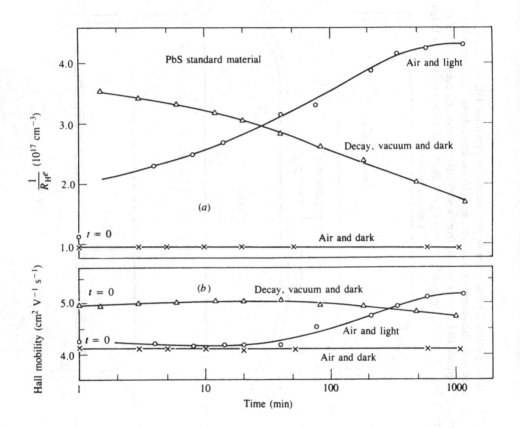

When the sample is exposed to air and light the mobility does not change for the first 40 min; for longer times the increase in conductivity is contributed equally by increases in carrier density and mobility. Similar behavior in reverse is shown to occur upon subsequent desorption of $O_2$ in vacuum and dark.

# 10

# Amorphous semiconductors

Although amorphous semiconductors have been known and investigated to some extent even longer than such well-known crystalline semiconductors as Si and GaAs that underlie the modern electronics revolution, many of their electronic and material properties have posed – and still do pose – an enigma for any kind of complete description. These amorphous semiconductors do not show the long-range order typical of crystalline materials with a periodic potential. As a consequence many of their properties differ significantly from their crystalline counterparts. The wavevector **k**, so important in describing electronic behavior in crystalline semiconductors, no longer has any significance. There is no density-of-states forbidden gap, but rather a continuous range of states throughout what was the forbidden gap in the corresponding crystalline material. Their conductivity cannot generally be increased appreciably by the incorporation of impurities, either because the random network of atoms can accommodate an impurity without leading to an extra electron or hole, or because the high density of localized states in the 'forbidden gap' effectively pins the Fermi level.

With the exception of amorphous Se, which has been widely used in electrophotography, most other amorphous semiconductors have not become serious competitors for electronic applications. This situation was dramatically changed by the discovery in the early 1970s by LeComber and Spear[1] that amorphous Si could be made to exhibit properties much more similar to those of a crystalline semiconductor if it were hydrogenated with concentrations of $H_2$ as high as 20% (a-Si:H). This has led to a large research and development effort with application of a-Si:H (and its Group IV solid solutions) to solar cells, electrophotography, thin film transistors, solid-state image sensors, optical recording, and a variety of amorphous junction devices.[2] Still a variety of problems unique to the amorphous state continues.

## 10.1 Background information on amorphous semiconductors

### Preparation

Amorphous materials can be made by deposition from the vapor phase, cooling from a liquid melt, or transformation of a crystalline phase by particle bombardment, oxidation etc. Amorphous materials that can be prepared by cooling from the melt are commonly called 'glasses.'

### Types of amorphous materials

There are three general categories of amorphous semiconductors: (*a*) covalent solids such as tetrahedral films of Group IV elements, or III–V materials, tetrahedral glasses formed from II–IV–V ternary materials such as $CdGeAs_2$, or chalcogenide glasses formed from the Group VI elements, or IV–V–VI binary and ternary materials; (*b*) oxide glasses such as $V_2O_5$–$P_2O_5$ that have ionic bonds and show electrical conductivity between different valence states of the transition metal ion ($V^{4+} \rightarrow V^{5+}$), (*c*) dielectric films such as $SiO_x$ and $Al_2O_3$. In this chapter we use the amorphous semiconductors associated with chalcogenides and a-Si:H to illustrate some of the properties and photoeffects.

### General description

Because of the lack of long-range order and the resulting high density of localized states in the 'forbidden gap,' the 'band picture' for amorphous semiconductors is different from that of their crystalline counterparts. An illustrative 'band' diagram is given in Figure 10.1.[3] This model includes (*a*) extended (non-localized) states below $E_v$ and above $E_c$, (b) a high density of localized 'band tail' states between $E_v$ and $E_B$, and between $E_A$ and $E_c$, due to the lack of long-range order, and (*c*) a high density of localized 'defect' states between $E_B$ and $E_A$.

Unlike the case in a crystalline material, the energies $E_v$ and $E_c$ do not indicate discontinuities in the density of states, but rather discontinuities in mobility (hence they are often called 'mobility edges'). In the extended states, the mobility ($1$–$10\ cm^2\ V^{-1}\ s^{-1}$) is much smaller than in a typical semiconductor material, but is much larger than in the localized states. Even in the extended states, the mean free path is shorter than the interatomic spacing, and carrier transport is primarily by diffusion.

Electrical conductivity can occur as a result of carrier motion in extended states, hopping conductivity via thermally-assisted tunneling between localized 'tail' states near the mobility edges, and tunneling between localized states near the Fermi level.

The effect of hydrogenation of amorphous Si is greatly to reduce the density of localized states in the 'forbidden gap,' thus making it possible to dope the material n- and p-type through the inclusion of donor or acceptor impurities. These localized states are associated primarily with Si 'dangling bonds' – Si bonds that do not have their valence requirements met. The decrease in the density of localized states also increases the carrier lifetime and diffusion length. Figure 10.2 shows the density of states in a-Si:H both in the undoped state and after doping with P donors. Two other properties of a-Si:H are indicated: (*a*) an increase in localized defect density caused by doping, and (*b*) an increase in localized defect density caused by illumination with light (the Staebler–Wronski effect), which is a degradation mechanism for device operation that can be reversibly annealed away. These effects are discussed further in this chapter.

Figure 10.1. Mott–Davis energy band model for an amorphous semiconductor, with extended states, localized tail states, and localized defect states.[3]

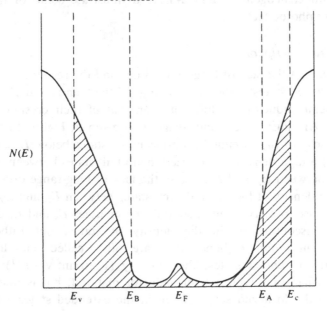

## 10.2 Amorphous chalcogenides

A wide variety of different amorphous chalcogenide semiconductors can be made by varying the composition of Group IV, Group V, and Group VI elements. A general pattern of behavior seems to characterize photoconductivity in most of these materials.[5]

### General photoconductivity characteristics

The temperature dependence of photoconductivity in a variety of different IV–V–VI chalcogenide semiconductors is illustrated in Figure 10.3. It is evident that the binary, ternary and more complex materials shown all exhibit approximately the same kind of photoconductivity behavior. These have been classified as showing Type I photoconductivity. Their photoconductivity has a maximum at a specific temperature $T_m$ with photoconductivity magnitude generally larger than the dark conductivity for $T < T_m$, and smaller than the dark conductivity for $T > T_m$.

Figure 10.2. Density of states for undoped (solid line) and P-doped (dashed line) a-Si:H.[4] For each case smaller variations due to illumination and annealing are shown.

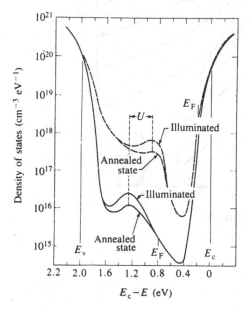

Figure 10.3. Temperature dependence of the dark conductivity and Type I photoconductivity for a variety of IV–V–VI chalcogenide semiconductors. (a) $Sb_2Te_3$ sputtered film, and (b) $As_2Te_3$ sputtered film, for photoexcitation by white light with maximum at 1.1 $\mu$m.[6] Highest light intensity $f^0$ corresponds to $2.5 \times 10^{17}$ photons $cm^{-2} s^{-1}$ of 1.1 $\mu$m radiation. Other $f$ superscripts indicate the optical density of the corresponding attenuation factor. (c) Sputtered $As_2SeTe_2$ film with same information as (a) and (b) except that $L_0 = 10^{18}$ $cm^{-2} s^{-1}$.[5] (d) Sputtered film of $Ge_{15}Te_{81}Sb_2S_2$.[7] $f = 1$ corresponds to $1.36 \times 10^{17}$ $cm^{-2} s^{-1}$. (e) Evaporated film of $Si_{11}Ge_{11}As_{35}P_3Te_{40}$, and (f) evaporated film of $Ge_{16}As_{35}Te_{28}S_{21}$, with measurement conditions similar to (d).[7]

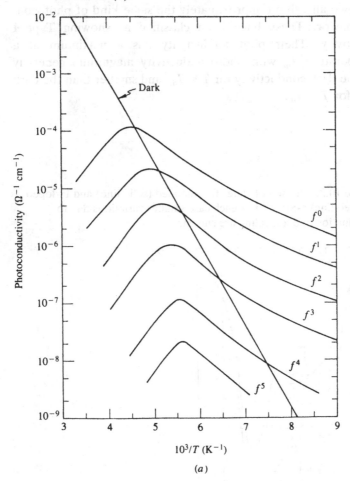

(a)

Quantitatively the photoconductivity behavior of Type I materials can be summarized as follows:

(1) For $T > T_m$, $\Delta\sigma < \sigma_d$, $\Delta\sigma \propto G$, $\Delta\sigma \propto \exp(E^+/kT)$.

(2) For $T \leq T_m$, $\Delta\sigma \propto G^{1/2}$ at high $G$, and $\Delta\sigma \propto G$ at low $G$, $\Delta\sigma \propto \exp(-E^-/kT)$.

(3) $T_m$ shifts to lower $T$ if $\Delta\sigma \propto G^{1/2}$, $T_m$ is independent of $T$ if $\Delta\sigma \propto G$.

(4) For $T \ll T_m$, $\Delta\sigma \gg \sigma_d$, $\Delta\sigma \propto G$, $\Delta\sigma$ approaches a constant value.

Some amorphous materials from the IV–V–VI family show a different kind of temperature dependence of photoconductivity that

Figure 10.3. (cont.)

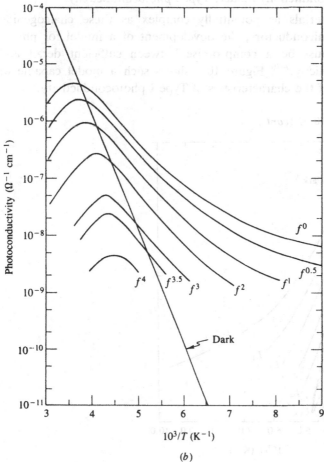

(b)

has been called Type II behavior: the photoconductivity maximum is absent, the photoconductivity simply increasing slowly and monotonically with increasing temperature, and, in general, the photoconductivity is much less than the dark conductivity. Examples are given in Figure 10.4. Definite Type II behavior is found in IV–V semiconductors such as $Sn_xAs_{1-x}$ and $Ge_xSb_{1-x}$, and in $Ge_xTe_{1-x}$ IV–VI materials for $x \geq 0.67$.

Several materials appear to be somewhat intermediate between Type I and Type II, such as $Ge_xAs_{1-x}$ ($x = 0.33$, $0.50$) and $Ge_xTe_{1-x}$ for $0.3 \leq x \leq 0.5$. For example, the $Ge_xAs_{1-x}$ materials exhibit a photoconductivity maximum and a well-defined $E^-$, but the magnitude of the photoconductivity is much less than the dark conductivity.

### Photoconductivity model: Type I photoconductivity

In materials as potentially complex as these chalcogenide amorphous semiconductors, the development of a model for photoconductivity must be a compromise between sufficient detail and necessary simplicity.[7–10] Figure 10.5 shows such a model capable of describing all of the characteristics of Type I photoconductivity.

Figure 10.3. (*cont.*)

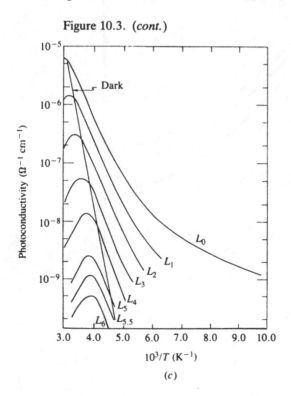

$(c)$

Figure 10.3. (*cont.*)

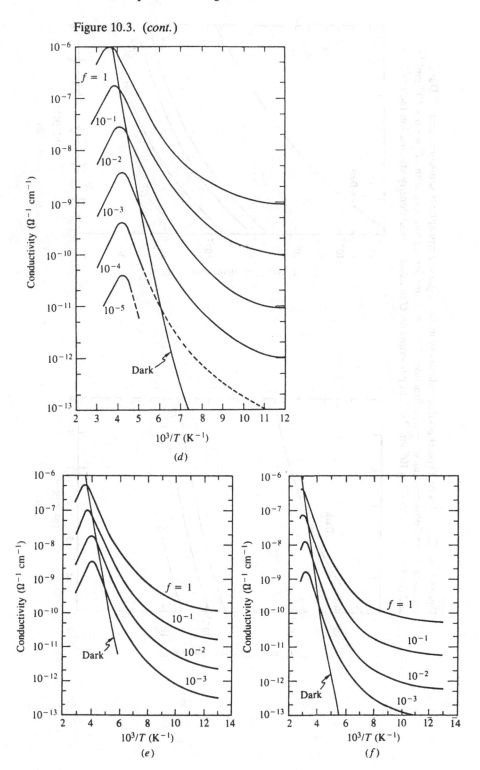

(d)

(e)

(f)

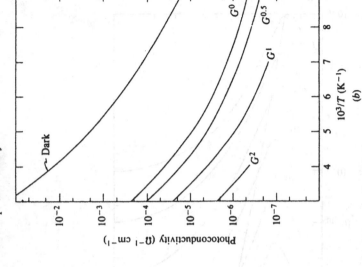

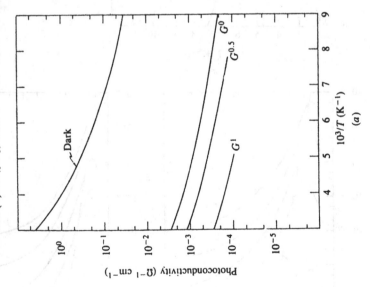

Figure 10.4. Examples of Type II photoconductivity in chalcogenide amorphous semiconductors.[11] Dark conductivity and photoconductivity as a function of temperature for (a) $Sn_{55}As_{45}$ with $G^0 = 2.6 \times 10^{20}$ cm$^{-3}$ s$^{-1}$ and (b) $Ge_{41}Sb_{59}$ with $G^0 = 10^{21}$ cm$^{-3}$ s$^{-1}$. Superscripts on $G$ indicate optical density attenuation factors.

Localized states extend into the gap from the conduction and valence edges. With increasing energy above the valence edge, the density of effective recombination centers decreases sharply as the energy exceeds $E_v^*$ for valence states, and increases sharply as the energy exceeds $E_c^*$ for conduction states.

Transitions 1 correspond to electron (hole) capture by a positively charged valence state below $E_v^*$ (negatively charged conduction state above $E_c^*$) with rate coefficient $C^+(C^-)$. Transitions 4 are similarly defined as capture of mobile electrons (holes) by a neutral conduction state above $E_c^*$ (neutral valence state below $E_v^*$) with rate coefficient $C^0$. Transitions 2 correspond to recombination between electrons localized in conduction states above $E_c^*$ and holes localized in valence states below $E_v^*$, with rate coefficient $K$. Transitions 3 represent recombination between an electron (hole) localized in a conduction state (valence state) above $E_c^*$ (below $E_v^*$) and holes (electrons) localized at energies near the equilibrium Fermi level, with rate coefficient $K'$ ($K''$). The equilibrium Fermi level lies $E_{F0}$ above the valence edge; since the conductivity is p-type, $E_{F0} \leq E_G/2$.

The formal procedure for calculating the steady state photoexcited

Figure 10.5. Proposed simple photoconductivity model for Type I photoconductivity in chalcogenide amorphous semiconductors.[7] (a) Schematic energy level diagram for the proposed model. (b) Typical recombination transitions considered in the photoconductivity model.

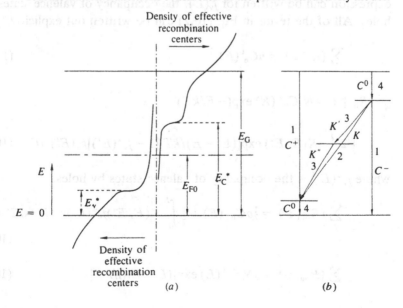

(a)                                                 (b)

carrier concentration is to (a) write the steady state occupancy function for a single localized level in terms of all possible transitions to other localized and non-localized states, and (b) write the free carrier continuity equation in terms of these occupancy functions integrated over all localized states.

As an example consider the rate at which electrons enter and leave a conduction state. In steady state these two rates must be equal. The occupancy function $f_c(e)$ for a conduction (C) state is given by

$$\sum_i (R_{in})_i [1 - f_c(E)] g_c(E) = \sum_j (R_{out})_j f_c(E) g_c(E) \tag{10.1}$$

where $(R_{in})_i$ is the product of the electron concentration in the initial state $i$ and the rate constant for an electronic transition from state $i$ to a $g_c(E)$ state; and $(R_{out})_j$ is the product of the concentration of available final states $j$ and the rate constant for an electronic transition from a $g_c(E)$ state to state $j$. Then

$$f_c(E) = \left[ \sum_i (R_{in} \downarrow)_i + \sum_i (R_{in} \uparrow)_i \right]$$

$$\times \left[ \sum_i (R_{in} \downarrow)_i + \sum_i (R_{in} \uparrow)_i + \sum_j (R_{out} \downarrow)_j + \sum_j (R_{out} \uparrow)_j \right]^{-1} \tag{10.2}$$

where we have separated the terms into upward ($\uparrow$) and downward ($\downarrow$) transitions of electrons into or out of the $g_c(E)$ state. A similar expression can be written for $f_v(E)$, the occupancy of valence states by holes. All of the terms in Eq. (10.2) can be written out explicitly.

$$\sum_i (R_{in} \downarrow)_i = n C_e^0(E) \tag{10.3}$$

$$\sum_i (R_{in} \uparrow)_i = N_v C_h^-(E) \exp(-E/kT)$$

$$+ 2 \int_0^E K(E, E') \exp[(E' - E)/kT][1 - f_v^*(E')] g_v(E') \, dE' \tag{10.4}$$

where $f_v^*(E')$ is the occupancy of valence states by holes.

$$\sum_j (R_{out} \downarrow)_j = \tfrac{1}{2} p C_h^-(E) + \tfrac{1}{2} \int_0^E K(E, E') f_v^*(E') g_v(E') \, dE' \tag{10.5}$$

$$\sum_j (R_{out} \uparrow)_j = \tfrac{1}{2} N_c C_e^0(E) \exp[(E - E_G)/kT] \tag{10.6}$$

The factors $\frac{1}{2}$ and 2 enter these equations as a result of assuming that the localized states are s-like with a spin degeneracy of 2, capable of being occupied by only one electronic charge.

To be explicit, the recombination coefficients have the following meaning: $C_e^+(E)$ is the capture coefficient for a non-localized electron by a hole in a valence level at $E$; $C_h^-(E)$ is the capture coefficient for a non-localized hole by an electron in a conduction level at $E$; $C_e^0(E)$ is the capture coefficient for a non-localized electron by an empty conduction level at $E$; $C_h^0(E)$ is the capture coefficient for a non-localized hole by an electron-occupied valence level at $E$; and $K(E, E')$ is the transition rate coefficient for an electron initially in a conduction state at $E$ recombining with a hole in a valence state at $E'$.

The continuity equation for non-localized electrons under steady state excitation $G$, the rate of generation of non-localized electron–hole pairs by incident light, is given by

$$dn/dt = 0 = G - \int_0^{E_G} nC_e^0(E)[1 - f_c(E)]g_c(E)\,dE$$

$$-\frac{1}{2}\int_0^{E_G} nC_e^+(E)f_v^*(E)g_v(E)\,dE$$

$$+\frac{1}{2}\int_0^{E_G} N_cC_e^0(E)\exp[(E - E_G)/kT]f_c(E)g_c(E)\,dE$$

$$+\int_0^{E_G} N_cC_e^+(E)\exp[(E-E_G)/kT][1-f_v^*(E)]g_v(E)\,dE$$

$$(10.7)$$

A continuity equation similar to Eq. (10.7) can be written for non-localized holes.

Expressions like those given in Eqs. (10.2) and (10.7) can in principle be used to calculate the steady state photoexcited carrier densities as a function of temperature and light intensity, provided that certain assumptions are made about the energy dependence of the densities of localized states and of the rate coefficients. In order to illustrate the physical significance of the consequence of such calculations, however, it is useful to make certain simplifying assumptions based on a careful evaluation of the significant features of the model.

Figure 10.5 has been drawn to illustrate one of these simplifications. In order to obtain an exponential variation of photoconductivity with $1/T$ above and below the photoconductivity maximum, it becomes clear that it is necessary to limit the distribution of effective localized states controlling recombination to regions of the mobility gap near the mobility edges. The actual shape of the localized state distributions is not critical, as long as there is a rapid change at $E_c^*$ and $E_v^*$, compared to changes at higher or lower energies; in the extreme case $E_c^*$ and $E_v^*$ could be approximated by discrete energy levels. We therefore consider a density of localized states $G_c$ at and above $E_c^*$, and a density of localized states $G_v$ at and below $E_v^*$. We assume, furthermore, that the mobility of non-localized carriers can be expressed as $\mu = \mu_0 \exp(-E_\mu/kT)$, where $E_\mu = E_\sigma - E_{F0}$, where $E_\sigma$ is the activation energy from $\ln \sigma$ vs $1/T$ plots, and we can obtain $E_{F0}$ from thermoelectric power data.

An evaluation of the total spectrum of possibilities afforded by this model leads to the conclusion that the following relationships are the only ones consistent with the model and its application to a variety of materials showing Type I photoconductivity. We assume that in the high-temperature range, transitions 1 or 2 dominate; in the intermediate-temperature range, transition 2 dominates at high intensities and transition 3 at low intensities; and in the low-temperature region, transition 4 dominates. Under these conditions fairly simple expressions can be derived for the temperature dependence of $\Delta\sigma$ in the high-temperature, intermediate-temperature, and low-temperature regimes, assuming $\Delta p \gg \Delta n$, as is the case for these materials.

In the high-temperature range, $\Delta p \ll p_0$, and it is formally possible for either transitions 1 or 2 to dominate. In either case, charge neutrality is controlled by states at the Fermi level. If transitions 1 dominate, then

$$\Delta\sigma_1 = (Gq\mu_0/2kTC_h^- G_c) \exp[(E_c^* - E_{F0} - E_\mu)/kT] \qquad (10.8)$$

If transitions 2 dominate, then

$$\Delta\sigma_2 = [Gq\mu_0 N_v/4(kT)^2 KG_c G_v] \exp[(E_c^* - E_v^* - E_{F0} - E_\mu)/kT]$$
$$(10.9)$$

General trends of observed behavior indicate that transitions 2 dominate at high temperatures as well as at high intensities at intermediate temperatures, although major differences in the analysis for densities and recombination coefficients do not result from using either Eq. (10.8) or (10.9) in place of the other with a typical material such as As$_2$SeTe$_2$.[10]

For intermediate temperature at high light intensities, $\Delta p \gg p_0$,

Table 10.1. *Typical values of parameters for* $As_2SeT_2$ *derived from photoconductivity measurements.*

| | |
|---|---|
| $E_{F0}{}^0 = 0.41$ eV | $\mu_{LT} = 10^{-3}$ cm$^2$ V$^{-1}$ s$^{-1}$ |
| $E_\mu = 0.14$ eV | $N_v = 1.7 \times 10^{20}$ cm$^{-3}$ |
| $E_v{}^{*0} = 0.11$ eV | $G_v = 4.6 \times 10^{18}$ cm$^{-3}$ eV$^{-1}$ |
| $E_c{}^{*0} = 0.86$ eV | $G_C = 3.9 \times 10^{20}$ cm$^{-3}$ s$^{-1}$ |
| $E^- = 0.25$ eV | $K = 4.2 \times 10^{-6}$ cm$^3$ s$^{-1}$ |
| $E^+ = 0.20$ eV | $N_{F0}K'' = 2.1 \times 10^6$ s$^{-1}$ |
| $E_G{}^0 = 1.20$ eV | |

transition 2 dominates, and neutrality is determined by localized states near the mobility edges.

$$\Delta\sigma_{hi} = (G/2K)^{1/2}(N_v q\mu_0/G_v kT)\exp[-(E_v{}^* + E_\mu)/kT] \quad (10.10)$$

At low light intensities in the same temperature range, $\Delta p \ll p_0$, transitions 3 dominate, and neutrality is determined by states at the Fermi level.

$$\Delta\sigma_{l0} = (Gq\mu_0 N_v/N_{F0}K''G_v kT)\exp[-(E_v{}^* + E_\mu)/kT] \quad (10.11)$$

where $N_{F0}$ is the density of states at the Fermi level.

Finally for low temperatures, $\Delta p \gg p_0$, transitions 4 dominate, and neutrality is determined by states near the mobility edges.

$$\Delta\sigma_{LT} = (Gq\mu_{LT}/C_h{}^0 G_v E_v{}^*) \quad (10.12)$$

where $\mu_{LT}$ is an asymptotic mobility reached at low temperatures.

From Eqs. (10.9)–(10.12), it follows that

$$E_v{}^{*0} = E^- - E_\mu \quad (10.13)$$

$$E_c{}^{*0} = E^+ + E^- + E_{F0}|_{T=0} \quad (10.14)$$

Here the superscript '0' indicates the value of the quantity at $T = 0$ K, and results from the fact that it is the value of the energy at $T = 0$ K that is determined from a measurement of activation energy if the energy varies linearly with temperature.

It is evident that these equations are consistent with the characteristics of Type I photoconductivity. If they are applied to a series of photoconductivity curves as a function of light intensity and temperature for a given amorphous material, values of $N_v$, $G_v$, $G_c$, $K$, $N_{F0}K''$, $E_v{}^{*0}$ and $E_c{}^{*0}$ can all be derived provided that values are assumed for $\mu_0$ and $C_h{}^0$. For Type I photoconductors, using $\mu_0 = 10$ cm$^2$ V$^{-1}$ s$^{-1}$ and $C_h{}^0 = 10^{-9}$ cm$^3$ s$^{-1}$, the values obtained for different materials of this type are approximately the same. Specific values obtained for $As_2SeTe_2$ are given in Table 10.1.

Without going into the question in detail, it may be pointed out that Type II photoconductivity can, in principle, be described within this same model with the additional assumption that for some reason transitions 3 dominate the photoconductivity over the whole measurable range without transforming to a transition-2-dominant regime.

### Scaling of parameters

The energy parameters of these materials are found to scale closely with the absorption gap $E_G^0$ regardless of the specific composition of the material.[11] A survey of some 20 amorphous semiconductors of IV–V–VI, V–VI, IV–V and IV–VI types (showing both Type I and Type II photoconductivity) shows that all of them fall with some small scatter about a line representing $E_\sigma = 0.46E_G^0$, as shown in Figure 10.6. Furthermore, data obtained on a smaller group of seven of these materials, on which complete photoconductivity and thermoelectric power data were available, show that $E_c^{*0} = 0.79E_G^0$, $E_{F0}^0 = 0.36E_G^0$, $E_v^{*0} = 0.11E_G^0$, $E_\sigma = 0.49E_G^0$, and consequently $E_\mu = 0.13E_G^0$, as shown in Figure 10.7.

Not only do these energy parameters scale with the absorption gap, but also the magnitude of the photoconductivity scales with this gap. Figure 10.8 shows the value of both the maximum photoconductivity observed at the highest light level used, and the value of the low-temperature asymptotic photoconductivity (about $10^{-3}$ that of $\Delta\sigma_m$) at this same light level for all 20 materials of Figure 10.6.

An understanding of this behavior can be obtained by recognizing that the maximum photoconductivity should be describable approximately by Eq. (10.10) with $T \approx T_m$. If all the preexponential factors of

Figure 10.6. Dark conductivity activation energy $E_\sigma$ as a function of absorption gap $E_G^0$ for 20 amorphous semiconductors: IV–V–VI (circles), V–VI(x), IV–V (triangles), and IV–VI (crosses).[11]

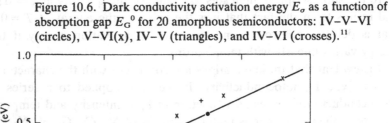
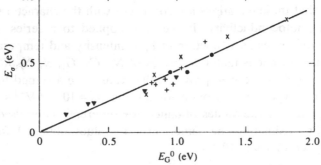

Figure 10.7. Energy parameters vs $E_G^0$ for seven amorphous semiconductor. The points represent, in order of increasing $E_G^0$: $Sn_{55}As_{45}$, $Sb_2Te_3$, $As_2Te_3$, $Ge_{15}Te_{81}Sb_2S_2$, $As_2SeTe_2$, $As_2Se_2Te$, and $As_2Se_3$.[11]

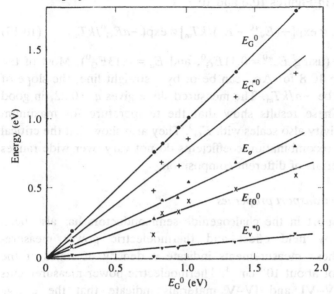

Figure 10.8. Magnitude of the photoconductivity at the maximum and at low temperatures as a function of absorption gap $E_G^0$, for 20 amorphous semiconductors: IV–V–VI (circles), V–VI (x), IV–V (triangles), and IV–VI (crosses).[11]

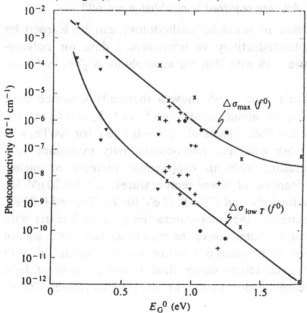

this equation were roughly the same for the different materials, then the major temperature dependence of the photoconductivity at $T_m$ would be given by the exponential term. Then, in view of the linear scaling shown in Figures 10.6 and 10.7,

$$\Delta\sigma_{max} \propto \exp[-(E_v{}^{*0} + E_\mu)/kT_m] \propto \exp(-aE_G{}^0/kT_m) \quad (10.15)$$

with $a = 0.24$ (using $E_v{}^{*0} = 0.11E_G{}^0$, and $E_\mu = 0.13E_G{}^0$). Most of the data in Figure 10.8 for $\Delta\sigma_m$ can be fit by a straight line, the slope of which should be $-a/kT_m$. The measured slope gives $a = 0.22$, in good agreement. These results show that the temperature for maximum photoconductivity also scales with $E_G{}^0$. They also show that the critical densities and recombination coefficients do not vary over wide ranges between materials of different composition.

### Other transport properties

Transport in the chalcogenide semiconductors has also been investigated by field effect and thermoelectric power measurements.[12,13] These measurements indicate a density of states at the Fermi levels of about $10^{19}$ cm$^{-3}$. Thermoelectric power measurements on V–VI, IV–VI and IV–V materials indicate that the charge transport is probably best described by a two-channel model with conduction simultaneously in extended states and localized states.

### How many data are required to establish a model?

A small lesson in scientific methodology can be learned by comparing the photoconductivity vs temperature data for polycrystalline PbS in Figure 9.19 with that for amorphous As$_2$Te$_3$ in Figure 10.3($b$).

Both materials are p-type. Both show a thermally activated dark conductivity with $\sigma_0$ of about $10\,\Omega^{-1}\,cm^{-1}$ and $E_\sigma = 0.42$ eV for As$_2$Te$_3$ and $0.35$ eV for PbS. Values of $E_\mu$ are $0.12$ eV for As$_2$Te$_3$ and $0.08$ eV for PbS. Both exhibit a photoconductivity maximum as a function of temperature with an exponential increase at higher temperatures and decrease at lower temperatures; $E^+ = 0.20$ eV for As$_2$Te$_3$ and $0.16$ eV for PbS, and $E^- = 0.17$ eV for As$_2$Te$_3$ and $0.03$ eV for PbS. In both materials the photoconductivity varies linearly with light intensity for temperatures above the maximum and as the square root of intensity for high intensities below the maximum. In both materials the photoconductivity decay time is independent of light intensity and increases exponentially at high temperatures ($E^* =$

0.26 eV in $As_2Te_3$ and 0.22 eV in PbS), and depends on light intensity and is independent of temperature at low temperatures.

Certainly, one might conclude, the mechanism for photoconductivity must be the same in both materials. Theories have been advanced as confirmed by much less experimental agreement than this. By having been 'on the inside' of both materials in Chapters 9 and 10, however, we have the unique position of knowing why the same model cannot apply to both materials.

## 10.3 Hydrogenated amorphous Si (a-Si:H)

Photoconductivity in a-Si:H is affected by at least three types of localized states: the conduction band tail states, the valence band tail states, and deep localized states in the gap associated with dangling bonds. A major difference with the model shown in Figure 10.5 for amorphous chalcogenides is that recombination between carriers in extended states and the deep localized defects due to dangling bonds is usually considered as the major recombination step, a recombination process neglected in the previous section.

Another major difference in the case of a-Si:H is that the defect associated with dangling bonds is considered to be a multivalent defect with three energy states. These three states are described as a negatively charged state, $D^-$, occupied by two electrons; a neutral state, $D^0$, occupied by one electron, which is therefore spin-active; and a positively charged state, $D^+$, electron unoccupied, as shown in Figure 10.9.

### Optical determination of defect densities

In high-quality, undoped a-Si:H, the density of dangling bond defects may be less than $10^{16}$ cm$^{-3}$, as indicated in Figure 10.2. It is important to know the density of these defects in order to model the behavior of the material. The measurement of such a low density of defects in a film of a-Si:H only one or a few micrometers thick, cannot be done by optical transmission, but requires more sensitive techniques.

One such technique is to use photoconductivity resulting from direct photoexcitation from these defects to the conduction levels as a measure of the absorption, and hence of the state density. Amplification techniques applied to ac photoconductivity can make even a small signal measurable. In order to avoid spurious effects due to a variation in electron lifetime with Fermi level position, the measurement is usually carried out under constant photocurrent conditions; hence, its name CPM, after constant photocurrent measurement. When the measured variable is the light intensity required to keep the photocurrent constant as the deep defect and tail states are optically scanned with a monochromator, variations in electron lifetime are cancelled out.

Another optical method has also been used to measure these small defect densities in thin films of material. It has become known as PDS, for photodeflection spectroscopy[4], and is illustrated in Figure 10.10. The density of the defects is measured by an indirect measurement of the small optical absorption of the material. An ac monochromatic

Figure 10.9. Proposed model for photoconductivity in a-Si:H involving trivalent recombination centers with states, $D^-$, $D^0$ and $D^+$, showing also the recombination coefficients $\beta$ and the various transitions included in the model. Also included are an exponential distribution of conduction- and valence-edge tail states, the occupancy of which is calculated solely in terms of the location of the dark or quasi-Fermi levels. Unoccupied energy levels are shown dashed.[15]

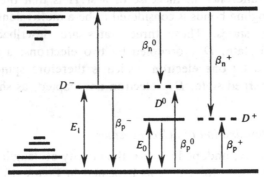

Figure 10.10. The experimental arrangement for defect determination in a-Si:H films by PDS.[4]

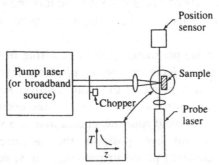

beam of light, corresponding to the photon energy for the energy difference between the defect states and the conduction states, is absorbed by the film which is surrounded by a suitable liquid (CCl$_4$ was initially used). Heat generated by the absorption flows into the liquid and produces an index of refraction gradient in the vicinity of the film surface. A weak laser beam is used to graze the surface of the film, where it is deflected by an amount proportional to the heat flow perpendicular to the sample surface. The deflection of the laser beam is measured with a position-sensitive detector.

### Effects of illumination on a-Si:H

A striking instability exists for a-Si:H when it is subjected to illumination, which manifests itself as a decrease in photoconductivity and in dark conductivity (as shown in Figure 10.11) due to an increase in the density of dangling bonds as illustrated in Figure 10.2.[14] These changes induced by illumination are stable at room temperature but can be reversed completely by thermal annealing at 200 °C. Related to these effects is an increase in the density of dangling bonds upon doping of a-Si:H, coupled with an increase in the density of additional dangling bonds created by photoexcitation of the doped material, as shown in Figure 10.2. The detailed properties and atomic description of these dangling bonds is a continuing area of research with many complex nuances and implications for practical devices involving a-Si:H.

### Photoconductivity model

Here we limit ourselves to considering a simple model of photoconductivity applicable to some extent, at least, to a-Si:H. We have already described the energy level scheme for this model briefly in Figure 10.9. In addition to the multivalent defect levels, $D^-$, $D^0$, and $D^+$, the model also includes exponential conduction band tail states with $N_{tn}(E) = 10^{21} \exp[-(E_c - E)/kT_n^*]$ with $kT_A^* = 0.025$ eV, and exponential valence band tail states with $N_{tp}(E) = 10^{21} \exp(-E/kT_p^*)$ with $kT_p^* = 0.045$ eV. We consider only capture transitions involving the recombination center and electrons or holes in extended states, and we calculate the density of trapped electrons or holes simply from the location of the dark or quasi-Fermi levels.

The model of Figure 10.9 includes thermal excitation of electrons from the valence band to the $D^+$ center at a rate $[D^+]N_v\beta_p^0 \exp(-E_0/kT)$; thermal excitation from the $D^-$ level to the conduction band at a rate $[D^-]N_c\beta_n^0 \exp[-(E_G - E_1)/kT]$; and thermal excitation of an electron from the valence band to the $D^0$ level at a rate $[D^0]N_v\beta_p^-\exp(-E_1/kT)$, where the $\beta$s are the appropriate recombination coefficients. Thermal excitation from the $D^0$ level to the conduction band has not been included in the current form of the model, but could easily be added if deemed necessary for comparison with results above room temperature.

The similarity with the one-center recombination model in Figure 3.5(e) is evident. Three general expectations may be stated simply from a consideration of this previous simple exponential trap model.

(1)   We can expect the value of $\gamma$ in $\Delta\sigma\propto G^\gamma$ to be controlled by occupancy of the conduction band tail states ($0.5\le\gamma\le1.0$ if

Figure 10.11. Decrease in the photoconductivity (solid line) and dark conductivity (dashed line) of an undoped a-Si:H film during illumination with 200 mW cm$^{-2}$ white light in the original experiments of Staebler and Wronski.[14]

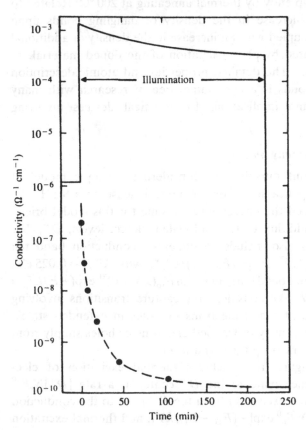

$T^* \geq T$) *only* when the photoexcited density of electron-occupied conduction band tail states is approximately equal to or at least proportional to the photoexcited density of $D^0$ (empty $D^-$) states. This condition requires that the dark densities of these two quantities be small, and that these two densities control the particle conservation relation. Specific expressions are given in the model below.

(2)  Even when these conditions are fulfilled, the measured value of $\gamma$ will be 0.50 because $[T^*/(T + T^*)] = 0.50$ for the specific borderline case of electron tail states in a-Si:H at 300 K where $kT^* = kT$.

(3)  If the density of $D^0$ centers is already appreciable in the dark, then the lifetime for low excitation rates will not vary with excitation rate, and $\gamma = 1.0$.

When the dark Fermi level lies close to the level corresponding to the $D^-$ centers (the condition that occurs for low doping or after optical degradation), a value of $\gamma \cong 1.0$ should be found at least for low excitation intensities. When the dark Fermi level lies well above the level of the $D^-$ centers (the condition that occurs for high n-type doping or after annealing of degradation), a value of $\gamma \cong 0.50$ should be found because of the specific value of $kT^* = kT$ for the conduction-band exponential tail distribution in a-Si:H.

Solution of the photoconductivity problem for a-Si:H proceeds along the lines developed in Section 3.3, with such variations as are required by the fact that the recombination center in this case is a trivalent defect.

In steady state five equations are needed in order to solve for the five unknowns, $n$, $p$, $[D^-]$, $[D^0]$, and $[D^+]$. The first equation is either the continuity equation for electrons,

$$0 = \partial n/\partial t = G - n[D^0]\beta_n{}^0 - n[D^+]\beta_n{}^+ + [D^-]P^- - np\beta_i$$

(10.16)

or the equivalent continuity equation for holes,

$$0 = \partial p/\partial t = G - [D^-]p\beta_p{}^- - [D^0]p\beta_p{}^0 + [D^+]P^+ + [D^0]P^0 - np\beta_i$$

(10.17)

Three more equations are obtained from the continuity equations for the three states of the dangling bond. For example, the continuity equation for $[D^-]$ is

$$0 = \partial[D^-]/\partial t = n[D^0]\beta_n{}^0 - [D^-]p\beta_p{}^- - [D^-]P^- + [D^0]P^0 \quad (10.18)$$

which gives the ratio,

$$[D^-]/[D^0] = (n\beta_n^0 + P^0)/(p\beta_p^- + P^-) \tag{10.19}$$

In the same way continuity equations for $[D^0]$ and $[D^+]$ can be solved for the ratio,

$$[D^0]/[D^+] = (n\beta_n^+ + P^+)/p\beta_p^0 \tag{10.20}$$

and then, with the recognition that the total density of dangling bonds $[D] = [D^-] + [D^0] + [D^+]$, the following equations are obtained to represent three of the five needed for solution of the problem.

$$[D^0] = [D]/Z \tag{10.21}$$

where

$$Z = [1 + (n\beta_n^0 + P^0)/(p\beta_p^- + P^-) + p\beta_p^0/(n\beta_n^+ + P^+)]$$

$$[D^-] = [(n\beta_n^0 + P^0)/(p\beta_p^- + P^-)][D]/Z \tag{10.22}$$

and

$$[D^+] = [p\beta_p^0/(n\beta_n^+ + P^+)][D]/Z \tag{10.23}$$

The fifth required equation can be obtained from the principle of particle conservation, namely that the increase in electron-occupied states above the dark Fermi level must be equal to the decrease in electron-occupied states below the dark Fermi level. Examples are given below.

The dark values of $[D^-]_0$, $[D^0]_0$, and $[D^+]_0$ can be calculated from a knowledge of the location of the dark Fermi level ($E_{F0}$ above the edge of the valence band) in the dark.

$$[D^-]_0/[D^0]_0 = \tfrac{1}{2}\exp[(E_{F0} - E_1)/kT] = A \tag{10.24}$$

$$[D^0]_0/[D^+]_0 = \tfrac{1}{2}\exp[(E_{F0} - E_0)/kT] = B \tag{10.25}$$

Thus if $[D]$ is the total density, and if $W = (AB + B + 1)$,

$$[D^-]_0 = AB[D]/W \tag{10.26}$$

$$[D^0]_0 = B[D]/W \tag{10.27}$$

$$[D^+]_0 = [D]/W \tag{10.28}$$

We also have

$$n_0 = N_c\exp[-(E_G - E_{F0})/kT] \tag{10.29}$$

$$p_0 = N_v\exp(-E_{F0}/kT) \tag{10.30}$$

To calculate the density of occupied electron traps in the dark, we assume that all electron (hole) traps below (above) the electron (hole) dark Fermi level are occupied.

$$n_{tn0} = kT_n^* N_{0n} \exp[-(E_c - E_{F0})/kT^*] \tag{10.31}$$

$$n_{tp0} = kT_p^* N_{0p} \exp(-E_{F0}/kT^*) \tag{10.32}$$

In the light we make the same assumption, but replace the dark Fermi level by the quasi-Fermi level.

$$n_{tn} = kT_n^* N_{0n}(n/N_c)^{(T/T^*)} \tag{10.33}$$

$$n_{tp} = kT_p^* N_{0p}(p/N_v)^{(T/T^*)} \tag{10.34}$$

Substitution of Eqs. (10.21)–(10.23), and (10.26)–(10.34), into a particle conservation equation allows the calculation of $p(n)$ or $n(p)$ for any desired temperature. Substitution of this result into Eq. (10.16) or (10.17) allows the calculation of $n(G)$ and $p(G)$.

We give the following three examples of the use of particle conservation.

(a)   If $(E_{F0} - E_1) \gg kT$

   Dark:   $[D^-]_0 \cong [D]; \ [D^0]_0 \gg [D^+]_0 \approx 0$

   Light:   $(n - n_0) + (n_{tn} - n_{tn0}) = \{[D^0] - [D^0]_0\}$
   $$+ 2\{[D^+] - [D^+]_0\} + (n_{tp} - n_{tp0}) + (p - p_0) \tag{10.35}$$

(b)   If $E_0 < E_{F0} < E_1$

   Dark:   $[D^0]_0 \cong [D]; \ [D^-] \approx [D^+]_0 \approx 0$

   Light:   $(n - n_0) + (n_{tn} - n_{tn0}) + \{[D^-] - [D^-]_0\}$
   $$= \{[D^+] - [D^+]_0\} + (n_{tp} - n_{tp0}) + (p - p_0) \tag{10.36}$$

(c)   If $E_{F0} < E_0$

   Dark:   $[D^+] \cong [D]; \ [D^0] \ll [D^-] \approx 0$

   Light:   $(n - n_0) + (n_{tn} - n_{tn0}) + 2\{[D^-] - [D^-]_0\}$
   $$+ \{[D^0] - [D^0]_0\} = (n_{tp} - n_{tp0}) + (p - p_0) \tag{10.37}$$

The numerical values assumed for the various parameters in the following calculations are summarized in Table 10.2. More or less idealized values have been used for capture coefficients.

*Doping of a-Si:H.*     By combining data from LeComber and Spear[16] and Street,[17] typical numbers were developed in which the

Table 10.2. *Values of parameters used in model calculations.*

| | |
|---|---|
| $E_1$ (eV) | 0.90 |
| $E_0$ (eV) | 0.45 |
| $N_c$ (cm$^{-3}$) | $3 \times 10^{19}$ |
| $N_v$ (cm$^{-3}$) | $1 \times 10^{19}$ |
| $N_{0n}$, $N_{0p}$ (cm$^{-3}$ eV$^{-1}$) | $1 \times 10^{21}$ |
| $v$ (cm s$^{-1}$) | $(3.03 \times 10^{11}T)^{1/2}$ |
| $\beta_p^-$ (cm$^3$ s$^{-1}$) | $1 \times 10^{-12}(300/T)^2 v$ |
| $\beta_p^0$ (cm$^3$ s$^{-1}$) | $1 \times 10^{-16} v$ |
| $\beta_n^0$ (cm$^3$ s$^{-1}$) | $1 \times 10^{-16} v$ |
| $\beta_n^+$ (cm$^{-3}$ s$^{-1}$) | $1 \times 10^{-12}(300/T)^2 v$ |
| $\beta_i$ (cm$^{-3}$ s$^{-1}$) | $1 \times 10^{-19} v$ |
| $\mu_n$ (cm$^2$ V$^{-1}$ s$^{-1}$) | 4–10 |

density of dangling bonds were assumed to vary with doping from $10^{16}$ cm$^{-3}$ for $E_{F0} = 0.85$ eV to $1.3 \times 10^{17}$ cm$^{-3}$ for $E_{F0} = 1.30$ eV in n-type material, and a symmetric behavior was simply assumed for p-type material with the density of dangling bonds increasing to $1.3 \times 10^{17}$ cm$^{-3}$ for $E_{F0} = 0.40$ eV. Calculations show that the actual

Figure 10.12. Variation of the photoconductivity in a-Si:H as a function of the location of the dark Fermi level as varied by doping.[15] The solid data points are from LeComber and Spear.[16] A mobility value of 10 cm$^2$ V$^{-1}$ s$^{-1}$ was assumed for calculation of the photoconductivity from the carrier density. Circles are for P doping, triangles for undoped material, and squares for B-doped material. The open circles and curve are the results of the model described here.

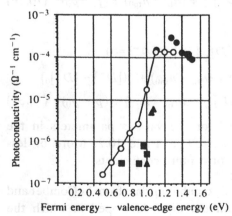

magnitude of the density of dangling bonds has much less effect on the behavior than the location of the dark Fermi level.

Figures 10.12 and 10.13 show the comparison of the calculated values of photoconductivity with the results of LeComber and Spear. There is a strong similarity, both qualitatively and quantitatively, between the predictions of the model and the data of LeComber and Spear, both for the variation of photoconductivity with location of the dark Fermi level, and for the exponent $\gamma$ in the power-law dependence of photoconductivity on excitation rate. These results are particularly striking, because, with the exception of the choice of $E_1 = 0.90$ eV and $\mu_n = 10$ cm$^2$ V$^{-1}$ s$^{-1}$, no other parameters were adjusted to provide the fit.

*Optical degradation.*　　Data available from Staebler and Wronski[14] may be used to assign values of the dark Fermi level position corresponding to various stages of optical degradation. Reasonable values of dangling bond densities are assumed from the data of Stutzman, Jackson, and Tsai[18] and Curtins *et al.*[19] The expected variation of photoconductivity with excitation rate may be calculated for four samples in various stages of optical degradation. The results of these calculations over a restricted range of excitation rates are given in Figure 10.14.

In the annealed state, the dark Fermi level lies 1.21 eV above the valence edge (somewhat higher than in undoped materials produced more recently), the density of dangling bonds is assumed to be $10^{16}$ cm$^{-3}$, and a value of $\gamma = 0.50$ is found over the whole range plotted. As optical degradation proceeds, the dark Fermi level moves

Figure 10.13. Variation of the value of $\gamma$ such that $\Delta\sigma \propto G^\gamma$, as a function of the location of the dark Fermi level as varied by doping.[15] The solid data points are taken from a curve from LeComber and Spear,[16] and the open circles and curve are the result of the model described here.

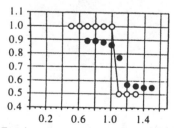

Fermi energy − valence-edge energy (eV)

down toward the $D^-$ level, a significant fraction of the $D^-$ levels are converted to $D^0$ levels in the dark, and $\gamma = 1.0$ at lower light levels for which the density of $D^0$ centers is essentially unchanged by the illumination.

The transition from $\gamma = 0.50$ to 1.0 shifts to higher excitation rates with increasing degradation. For the most degraded sample used, the dark Fermi level lies 1.00 eV above the valence edge, the density of dangling bonds is assumed to be $10^{17}\,\mathrm{cm}^{-3}$, and the value of $\gamma = 1.0$ over this whole range.

The similarities between the curves of Figure 10.14 and the experimental data reported by Staebler and Wronski indicate that the major effects can be explained as the consequence of a decrease in the value of the dark Fermi energy caused by the increase in the total density of dangling bonds created by photoexcitation. Both the data

Figure 10.14. Photoexcited electron density as a function of photoexcitation rate for a sample of a-Si:H at four stages of optical degradation (corresponding to values of $E_{F0} = 1.21$, 1.09, 1.04 and 1.00 eV from top to bottom).[15] The highest data are for an annealed sample, and the lowest data are for a degraded sample corresponding to 4 h of illumination with 200 mW cm$^{-2}$ of red light. Solid data points are inferred from the data of Staebler and Wronski[14] with photoconductivity converted to electron density using a mobility of 4 cm$^2$ V$^{-1}$ s$^{-1}$ and with the excitation rate selected so that $10^{17}$ photons cm$^{-2}$ s$^{-1}$ is taken to be an excitation rate of $10^{20}$ cm$^{-3}$ s$^{-1}$. They correspond to the following values of dark Fermi level (eV), and assumed dangling-bond density (cm$^{-3}$): (circles) 1.21, $10^{16}$; (squares) 1.09, $6 \times 10^{16}$; (triangles) 1.04, $8 \times 10^{16}$; (circles) 1.00, $10^{17}$ (top to bottom) used for the curves calculated from the model described here.

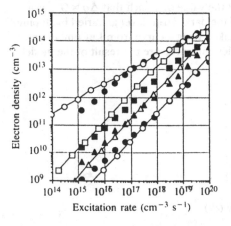

and the calculations of Figure 10.14 indicate that the decrease in photoconductivity at lower excitation rates with optical degradation is appreciably greater than would be expected from a simple increase in the density of dangling-bond recombination centers by an order of magnitude.

It is evident that remaining to be explained is the cause of the discrepancy between the value of $\gamma = 1.0$ obtained from this simple model and the value of $\gamma = 0.8$–$0.9$ found experimentally at lower light levels, as shown in Figure 10.14. Other levels or transitions may well be involved so that the electron lifetime decreases with increasing excitation rate even in a regime dominated by a major fraction of the dangling bonds existing in the $D^0$ state in the dark.

# 11

# Photovoltaic effects

In most of our discussion to this point, we have been dealing with a change in electronic properties of a semiconductor with an externally applied electric field when it was subjected to illumination. If the semiconductor has an internal electric field, then the electron–hole pairs generated by optical absorption can be separated by this internal field and give rise to a voltage and/or a current in an external circuit. In this way incident radiation can be converted into electrical power.

Many books and reviews have been written on this subject,[1–4] and it is not our purpose here to attempt to give any kind of exhaustive treatment of photovoltaic materials, effects and devices, or solar cell design. Our goal is much more modest: to provide a brief overview of the major properties of different kinds of photovoltaic systems, sufficient to enable us to consider the specific physical phenomena that can be observed in photovoltaic systems, and to give a few illustrative examples.

The specific photovoltaic system involving $Cu_xS/CdS$ heterojunctions was among the first to be developed for thin-film solar cells, but its susceptibility to degradation mechanisms has led to its essential abandonment as a commercial candidate. Nevertheless for our purposes here, it serves as a rich 'theater for photoelectronic effects,' most of which have been mentioned earlier in this book, providing illustrations of many of these effects in specific form.

## 11.1 Types of semiconductor junctions

There are four principal ways to obtain the internal field in a semiconductor structure needed to produce photovoltaic behavior: homojunctions, heterojunctions, buried or heteroface junctions, and

Schottky barriers. In addition, some particular materials are more suited for use in a p–i–n device, where an undoped insulating region makes up most of the device with high-conductivity p- and n-type regions.

### Homojunctions

We have previously discussed p–n junctions briefly as far as their use as photodetectors is concerned. In this application the junction is reverse biased and the increase in reverse saturation current with illumination is measured. When a homojunction is used as a solar cell, it does not have any applied bias. A representative energy band diagram is given in Figure 11.1; in this diagram a symmetric distribution of the depletion region between n- and p-type regions is shown, such as would be found if the carrier density in both

Figure 11.1. Energy band diagram for a p–n homojunction in a semiconductor, showing the vacuum level $E_{vac}$, the conduction band edge $E_c$, the Fermi level $E_F$, and the valence band edge $E_v$, as well as the band gap $E_G$, the electron affinity $\chi_s$, and the diffusion potential $\phi_D$.[2] The inset shows a typical geometry for a photovoltaic cell, with a narrow n-type region on a wider p-type region.

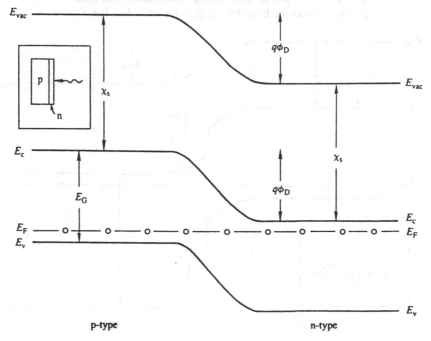

n- and p-type regions were the same. The most common Si p-n junction solar cells are homojunctions.

### Heterojunctions

Heterojunctions are p–n junctions in which the p-type material is different from the n-type material. An idealized energy band diagram for a heterojunction is given in Figure 11.2, following the Anderson abrupt junction model[5] that neglects any effects due to interface states and dipoles. This model includes the possibility of energy discontinuities in the conduction band and the valence band at the junction interface because of differences in electron affinity and bandgap between the two semiconductors. Figure 11.2 has been plotted with a choice of $E_G$s and $\chi_s$s such that the discontinuities in both bands are downwards. If $\chi_1$ had been chosen to be larger than $\chi_2$, then an energy spike (an upward discontinuity) would have been formed at the conduction band, which would have seriously impeded

Figure 11.2. Energy band diagram for a p–n heterojunction between two different semiconductors, with material parameters such that there are no energy spikes at the junction interface.[2] The inset shows two possible modes of operation: 'front-wall' with illumination on the small bandgap, p-type absorbing semiconductor, or 'back-wall' with illumination on the large bandgap, n-type window layer.

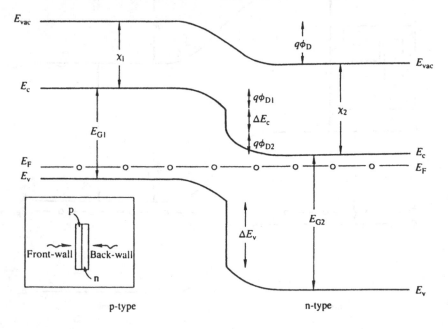

carrier flow over the junction. Also in Figure 11.2 the p-type material has been chosen to have the smaller bandgap, a common choice since electron diffusion lengths in p-type materials are usually larger than hole diffusion lengths in n-type materials. By having a large band n-type window material, such heterojunctions allow light to be transmitted through the window layer to be absorbed in the p-type material in the immediate vicinity of the p–n junction.

If the two materials making up the heterojunction do not have matching lattice spacings, then the existence of interface states at the junction occurs with possible consequences for the current flow processes controlling the solar cell behavior, as we describe below. Sometimes a variation on the *p–n* heterojunction (sometimes called an SS junction, where 'S' stands for 'semiconductor') is made by including a thin layer of an insulating material between the two semiconductors (an SIS junction) to help reduce the junction currents that decrease the magnitude of the voltage that can be developed under illumination.

### Buried or heteroface junctions

Figure 11.3 shows a representative energy band diagram for a 'buried' p–n homojunction with a heavily-doped $p^+$–p heterojunction on the illuminated surface, so that the total structure is actually a $p^+$–p–n junction. This structure retains the advantages of a p–n junction (e.g., absence of junction interface state effects), while at the same time providing a different front surface interface for the p-type region in the homojunction, with the possibility of decreasing surface losses. The typical example of the success of such a process is given by the AlGaAs/GaAs solar cell[6] in which the larger bandgap AlGaAs is the $p^+$ heteroface layer on a p–n GaAs junction. The lattice constant of AlGaAs almost exactly matches that of GaAs, making it possible to produce a heteroface interface with minimized density of interface states.

### Schottky barriers

Perhaps the conceptually simplest of the junction configurations useful for photovoltaic effects is the Schottky barrier (or MS junction) between a suitable metal and a semiconductor. A typical energy band diagram is given in Figure 11.4. Although a simple approach would equate the diffusion potential of the Schottky barrier to the difference between the work function of the metal and the semiconductor, many more complex interactions can occur at the metal–semiconductor interface that cause variations on this prediction. The actual height of a specific Schottky barrier must often be determined experimentally.

### p–i–n junctions

In photovoltaic applications of all of the above junctions, carrier diffusion to the junction is the principal mechanism for collection of carriers generated by light. Carriers photoexcited within the depletion region are also collected by the drift field of the depletion region, and are usually considered to a first approximation to be totally collected. Barring unforeseen difficulties, a deliberately chosen structure of p–i–n type, in which absorption of light takes place primarily in the i region with subsequent collection of all photoexcited carriers by drift in the electric field in the i region, appears to offer definite advantages.

An energy band diagram for a p–i–n junction in a-Si:H is shown in Figure 11.5. The diffusion lengths in doped a-Si:H are so low that attempts to use conventional junction forms in a-Si:H solar cells would not be successful. In efficient a-Si:H solar cells the electric field

Figure 11.3. Energy band diagram for a $p^+$–p–n heteroface buried junction in which the $p^+$ material acts as a large bandgap window and an ohmic contact to the p-type material.[2] Inset shows the typical direction of illumination for use as a solar cell.

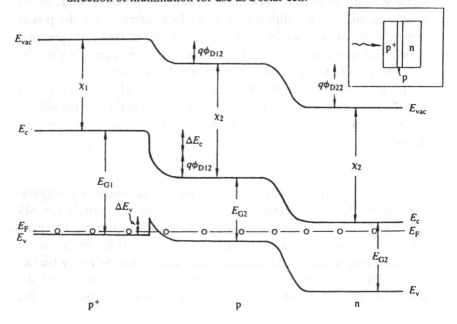

Figure 11.4. Energy band diagram for a Schottky barrier on an
n-type semiconductor, based on the energy parameters of the
materials without inclusion of interface effects.[2] The diagram shows
the work functions of the metal $q\phi_M$ and of the semiconductor $q\phi_s$.

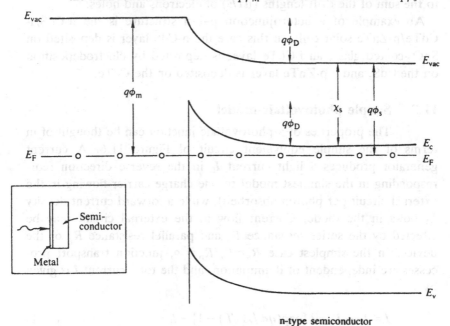

n-type semiconductor

Figure 11.5. An energy band diagram for a p–i–n cell of a-Si:H
pictured in the short-circuit mode with a $SnO_2$ transparent,
conducting coating.[7]

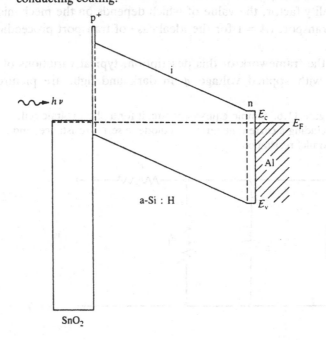

in the i region is almost uniform and the photocurrent is proportional to the sum of the drift lengths ($\mu\tau E$) of electrons and holes.

An example of a heterojunction p–i–n structure is the n-CdS/i-CdTe/p-ZnTe solar cell;[8] in this case the n-CdS layer is deposited on SnO$_2$-coated glass, an i-CdTe layer is deposited by electrodeposition on the CdS, and a p-ZnTe layer is deposited on the CdTe.

## 11.2    Simple photovoltaic model

The properties of a photovoltaic junction can be thought of in terms of the simple equivalent circuit of Figure 11.6. A current generator produces a light current $I_L$ in the reverse direction (corresponding in the simplest model to one charge carrier flowing in the external circuit per photon absorbed), while a forward current density $I_D$ flows in the diode. Current flow in the external circuit may be affected by the series resistance $R_s$ and parallel resistance $R_p$ of the device. In the simplest case $R_s \approx 0$, $R_p \approx \infty$, junction transport processes are independent of illumination, and the total current $I$ is given by

$$I = I_D + I_L = I_0[\exp(q\phi/AkT) - 1] - I_L \qquad (11.1)$$

where $I_0$ is the reverse saturation current of the diode (which depends on the actual transport mechanism for the diode current), and $A$ is the diode ideality factor, the value of which depends on the mechanism of junction transport ($A = 1$ for the ideal case of transport proceeding by diffusion).

Within the framework of this description, typical variations of total current $I$ with applied voltage $\phi$ in dark and light are pictured in

Figure 11.6. Simple equivalent circuit for a photovoltaic cell, including a current generator, a diode, a series resistance, and a parallel resistance.[2]

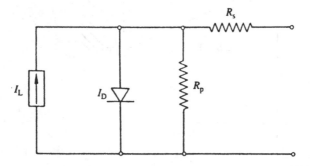

Figure 11.7. If the applied voltage is equal to zero, then

$$I_{sc} = -I_L \tag{11.2}$$

where $I_{sc}$ is the short-circuit current. The short-circuit current is controlled only by the current generation and recombination processes. If the current flowing $I = 0$, then

$$\phi_{oc} = (AkT/q) \ln[(I_L/I_0) + 1] \tag{11.3}$$

where $\phi_{oc}$ is the open-circuit voltage. The open-circuit voltage is controlled also by the diode current with its dependence on $A$ and $I_0$.

From Eqs. (11.2) and (11.3) we also see a relationship between $I_{sc}$ and $\phi_{oc}$ for this simple model:

$$I_{sc} = I_0[\exp(q\phi_{oc}/AkT) - 1] \tag{11.4}$$

which is exactly the same form as for the diode current vs voltage in the dark. Thus a plot of current vs voltage in the dark should have the same defining values of $A$ and $I_0$ as a plot of $I_{sc}$ vs $\phi_{oc}$.

Figure 11.7. Typical idealized light and dark current vs voltage curves for a photovoltaic cell, showing the open-circuit voltage, the short-circuit current, and the maximum power point.[2]

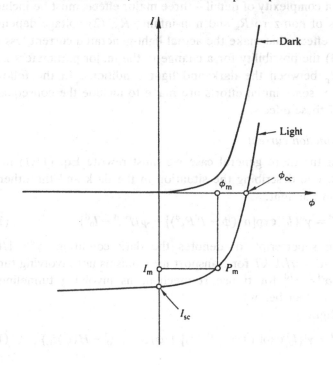

Since a practical solar cell is a power conversion device, it operates at maximum efficiency when its output product of current and voltage is a maximum. the efficiency $\eta$ can be expressed as

$$\eta = P_m/P_{rad} = I_m\phi_m/P_{rad} = I_{sc}\phi_{oc}ff/P_{rad} \tag{11.5}$$

where $P_{rad}$ is the total radiation power incident on the cell, and $ff$ is called the fill factor; $ff$ is a measure of the 'squareness' of the light $I-\phi$ curve. According to Eq. (11.5) its definition is

$$ff = I_m\phi_m/I_{sc}\phi_{oc} \tag{11.6}$$

The value of $\phi_m$ can be obtained by multiplying Eq. (11.1) by $\phi$ and then maximizing the power with respect to $\phi$. The result is

$$\phi_m = \phi_{oc} - (AkT/q) \ln[(q\phi_m/Akt) + 1] \tag{11.7}$$

which can be solved iteratively for $\phi_m$. The current for maximum power $I_m$ and hence the maximum power $P_m = I_m\phi_m$ can then be obtained from Eq. (11.1) by substituting $\phi = \phi_m$.

## 11.3    A more realistic photovoltaic model

In making the first step from the idealized model of the previous section to a more realistic model – without being enmeshed in too great a complexity of detail – three major effects must be included: (1) effects of non-zero $R_s$ and non-infinite $R_p$, (2) voltage-dependent collection effects that make the actual light-generated current less than $I_L$, and (3) the possibility for a change in the major parameters $I_0$, $A$, $R_s$ and $R_p$ between the dark and light conditions. In the following description, some initial efforts are made to include the consequences of each of these effects.

### Junction current

In the more general case we must rewrite Eq. (11.1) in two equations, one describing the situation in the dark and the other the situation in the light.

$$I^d = \gamma^d\{I_0^d \exp[\alpha^d(\phi - I^dR_s^d)] + \phi/R_p^d - I_0^d\} \tag{11.8}$$

where the superscript 'd' denotes the dark condition, $\gamma^d = 1/(1 + R_s^d/R_p^d)$, $\alpha^d = q/A^dkT$ for transport mechanisms not involving tunneling and $\alpha^d = \alpha'^d$ for transport mechanisms involving tunneling (as described further below).

In the light

$$I^l = \gamma^l\{I_0^l \exp[\alpha^l(\phi - I^lR_s^l)] + \phi/R_p^l - I_0^l - H(\phi)I_L\} \tag{11.9}$$

where $H(\phi)$ is a voltage-dependent collection function describing the fraction of the light-generated carriers that actually contribute to the light-generated current. Such functions may be approximated as follows.

### Collection functions

The use of a voltage-dependent collection function $H(\phi)$ is a relatively simple way to deal with the fact that the actual collected current may be less than the ideal light current $I_L$. For the sake of simplicity, consider a p–n heterojunction where the n-type material is a large bandgap window that light passes through without loss to be absorbed in the p-type material. The collection function may be divided into two separate contributions:

$$H(\phi) = g(\phi)h(\phi) \tag{11.10}$$

where $g(\phi)$ describes the loss of carriers by recombination in the bulk of the p-type material beyond the depletion region before they can diffuse to the junction to be collected, and $h(\phi)$ describes the loss of carriers by recombination at the junction interface due to interface states. It is expected that $g(\phi)$ will vary strongly with photon energy due to the variation of absorption constant with photon energy, and the greater probability for loss by recombination before diffusion to the junction for carriers treated at greater distances from the junction. On the other hand it is expected that $h(\phi)$ will be relatively independent of photon energy.

The collection function $g(\phi)$ can be calculated as follows:

$$g(\phi) = \left\{ \int_0^w \exp(-\alpha x)\, dx \right.$$

$$\left. + \int_w^\infty \exp(-\alpha x) \exp[-(x-w)/L_n]\, dx \right\} \Big/ \int_0^\infty \exp(-\alpha x)\, dx \tag{11.11}$$

where the dependence on photon energy occurs through the dependence of the absorption constant $\alpha(\hbar\omega)$. The first term in Eq. (11.11) expresses complete collection of carriers created in the depletion layer with width $w$, assuming that the drift field there exists in this collection. The second term expresses the possibility of recombination loss of such carriers with increasing distance from the depletion layer with width $w$ if the diffusion length of electrons in the p-type material is $L_n$. Integration of Eq. (11.11) gives

$$g(\phi) = 1 - \exp[-\alpha w(\phi)]/(1 + \alpha L_n) \tag{11.12}$$

where the dependence on the voltage $\phi$ through the variation of depletion width with $\phi$ is emphasized, as described by Eq. (1.24).

A simple form for the interface collection function $h(\phi)$ can be obtained by assuming that recombination at the interface is described by an interface recombination velocity $s_I$, and that the recombination probability can be considered to be a simple competition between crossing the junction without recombination and recombination at the interface.

$$h(\phi) = 1/(1 + s_I/\mu\mathscr{E}) \tag{11.13}$$

where $\mu$ is the mobility of carriers at the interface, and $\mathscr{E}$ is the electric field at the interface, given by $\mathscr{E} = 2(\phi_D - \phi)/w(\phi)$. If there are $N_I$ interface states per square centimeter at the interface with a capture coefficient of $\beta_I \, cm^3 \, s^{-1}$, then $s_I = N_I\beta_I$. The carrier velocity $\mu\mathscr{E}$ must be considered to have a maximum value corresponding to the saturation of drift velocity at high fields.

An illustration of these collection functions in action is given in the spectral response of quantum efficiency (electrons in external circuit divided by photons absorbed) for an n-CdS/p-CdTe heterojunction in Figure 11.8. In such a heterojunction, the short-wavelength cutoff is

Figure 11.8. Spectral dependence of the quantum efficiency for an n-CdS/p-CdTe heterojunction, illustrating the contributions of collection functions $g(\phi)$ and $h(\phi)$.[9]

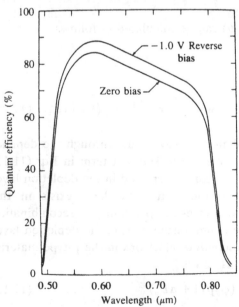

due to the absorption of shorter wavelengths by the large bandgap n-type CdS, while the long-wavelength cutoff is due to failure of the small bandgap p-type CdTe to absorb for longer wavelengths. The collection function $g(\phi)$ manifests itself in the negative slope of the quantum efficiency toward longer wavelengths, with negligible dependence on voltage $\phi$ over the range measured. The collection function $h(\phi)$ manifests itself in an increase in quantum efficiency with applied reverse bias (thus increasing $\mathcal{E}$) that is independent of wavelength, from a value $h(\phi) = 0.84$ at $\phi = 0$ to $h(\phi) = 0.89$ for $\phi = -1$ V. A value of $s_\text{I} = 2 \times 10^6$ cm s$^{-1}$ is consistent with all junction parameters for this case, which is in keeping with the fact that CdS and CdTe have a large 9% lattice mismatch.

### Other photovoltaic parameters

When the more realistic junction model is considered, then other characteristic photovoltaic parameters must be rewritten.

The open-circuit voltage now becomes

$$\phi_\text{oc} = (1/\alpha^\text{l}) \ln[H(\phi_\text{oc})(I_\text{L}/I_0^\text{l}) + 1 - \phi_\text{oc}/I_0^\text{l}R_\text{p}^\text{l}] \tag{11.14}$$

The short-circuit current becomes

$$I_\text{sc} = \gamma^\text{l}[I_0^\text{l} \exp(-\alpha^\text{l}I_\text{sc}R_\text{s}^\text{l}) - I_0^\text{l} - H(0)I_\text{L}] \tag{11.15}$$

The relationship between the open-circuit voltage and the short-circuit current becomes

$$-I_\text{sc} = \gamma^\text{l}[H(0)/H(\phi_\text{oc})][I_0^\text{l} \exp(\alpha^\text{l}\phi_\text{oc}) + \phi_\text{oc}/R_\text{p}^\text{l} - I_0^\text{l}]$$
$$+ \gamma^\text{l}I_0^\text{l}[1 - \exp(-\alpha^\text{l}I_\text{sc}R_\text{s}^\text{l})] \tag{11.16}$$

In order for this more realistic expression to satisfy the same requirements as those of Eq. (11.4), i.e., in order for the $I_\text{sc}$ vs $\phi_\text{oc}$ plot to show the same dependence as the $I^\text{d}$ vs $\phi$ plot, note that seven conditions must be met: $\gamma$, $R_\text{p}$, $\alpha$, and $I_0$ must all be independent of illumination; $H(0)$ must be equal to $H(\phi_\text{oc})$; $I^\text{d}R_\text{s}^\text{d}$ must be much less than $\phi$; and $\exp(-\alpha^\text{l}I_\text{sc}R_\text{s}^\text{l})$ must be of order unity.

In order to express the voltage $\phi_\text{m}$ and current $I_\text{m}$ at maximum power as in Eq. (11.7), the following two equations must be solved simultaneously:

$$\phi_\text{m} = (1/\alpha^\text{l}) \ln\{1 + (I_\text{L}/I_0^\text{l})[H(\phi_\text{m}) + \phi_\text{m}(\partial H/\partial \phi)_\text{m}] - 2\phi_\text{m}/I_0^\text{l}R_\text{p}^\text{l}\}$$
$$- (1/\alpha^\text{l}) \ln\{1 + \alpha^\text{l}[1 - R_\text{s}^\text{l}(\partial I^\text{l}/\partial \phi)_\text{m}\phi_\text{m}] + I_\text{m}^\text{l}R_\text{s}^\text{l} \tag{11.17a}$$

$$I_\text{m}^\text{l} = \gamma^\text{l}\{I_0^\text{l} \exp[\alpha^\text{l}(\phi_\text{m} - I_\text{m}^\text{l}R_\text{s}^\text{l})] + \phi_\text{m}/R_\text{p}^\text{l} - I_0^\text{l} - H(\phi_\text{m})I_\text{L}\} \tag{11.17b}$$

## 11.4    Junction transport processes in heterojunctions

It is not the purpose of this discussion to provide a detailed investigation of the many junction design decisions that are involved in optimizing a photovoltaic cell; such a discussion we leave to the more specific descriptions to which we have already referred. It is worthwhile, however, to point out the different kinds of processes that can contribute to the important parameters, $A$ and $I_0$, which act as ultimate limits on the magnitude of the open-circuit voltage of such a junction. A simple summary of the major junction transport processes is given in Figure 11.9. We use the example of an $n^+$–p heterojunction for the specific discussions following.

Figure 11.9. Schematic illustration of three major modes of forward junction current: (1) injection over the barrier, (2) recombination in the depletion layer, and (3) tunneling, with or without thermal assistance, through interface or imperfection states, followed by recombination.[2]

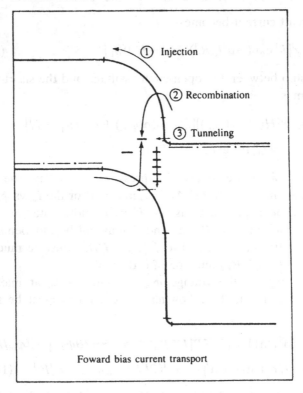

① Injection

② Recombination

③ Tunneling

Foward bias current transport

*Diffusion*

The most ideal behavior, producing the smallest values of $I_0$, corresponds to a diffusion-controlled current over the junction barrier, associated with the injection of electrons from the n-type material into the p-type material of Figure 11.9. After injection and diffusion, recombination finally occurs away from the junction in the p-type semiconductor bulk. Diffusion currents are generally much smaller than the forward transport currents observed in heterojunctions. They may be described by $I_{diff} = I_0[\exp(q\phi/kT) - 1]$, with $\alpha_{diff} = q/kT$ and $A = 1$, and

$$I_0 = (n_i^2/N_A)q(L_n/\tau_n) \qquad (11.18)$$

for the specific case of a p–n heterojunction with $N_D > N$, and $E_{Gp} < E_{Gn}$, as for example is encountered with a high conductivity, wide bandgap n-type layer on a p-type absorber material. A plot of $\ln(I_0 T^{-7/2})$ vs $1/T$ has an activation energy of $E_{Gp}$.

*Recombination in the depletion region*

The second most desirable junction transport mechanism corresponds to current transport through the junction through recombination in the depletion region. Such recombination currents are also usually smaller than the forward transport currents observed in heterojunctions. A recombination rate expression is used based on Shockley–Read recombination, with integration across the depletion regions. Such currents may be described by $I_{rec} = I_0[\exp(q\phi/2kT) - 1]$, with $\alpha_{rec} = q/2kT$ and $A = 2$, and

$$I_0 = qn_i[1/(\tau_{n0}\tau_{p0})^{1/2}][\pi kT/4(\phi_D - \phi)]w \qquad (11.19)$$

where $w$ is the depletion layer width. The value $A = 2$ is strictly true only if the recombination levels lie at midgap; in some practical situations $A$ may take on values between 1 and 2 and still be described in essentially this way. A plot of $\ln(I_0 T^{-5/2})$ vs $1/T$ has an activation energy of $E_{Gp}/2$.

*Interface recombination without tunneling*

At an n$^+$–p interface, electrons are abundant and current is limited by the supply of holes, which must surmount the barrier in the valence band. Such currents may be described by $I_{int} = I_0[\exp(q\phi/kT) - 1]$, with $\alpha_{int} = q/kT$ and $A = 1$.

If the thermal velocity $v_{th}$ of an electron is larger than the interface recombination velocity $s_I$, the current is limited by interfacial recombination:

$$I_0 = qs_I N_v \exp(-q\phi_D/kT) = qN_I S_I v_{th} N_v \exp(-q\phi_D/kT) \tag{11.20}$$

where $S_I$ is the capture cross section of interface states, and a plot of $\ln I_0$ vs $1/T$ has an activation energy of $q\phi_D$.

If the thermal velocity of an electron is smaller than the interface recombination velocity, the current is limited by diffusion of holes to the interface:

$$I_0 = A'T^2 \exp(-q\phi_D/kT) = (4\pi q m_h^*/h^3)(kT)^2 \exp(-q\phi_D/kT) \tag{11.21}$$

and a plot of $\ln(I_0 T^{-2})$ vs $1/T$ has an activation energy of $q\phi_D$.

### Tunneling limited recombination through interface states without thermal assistance

Electrons from the n-type material descend through interface states and then tunnel through the base of a barrier of height $E_b$ into the valence band of the p-type material. A simple model for this process involves tunneling at the base of a parabolic barrier.[10] The junction current $I_{ti} = I_0[\exp(\alpha/kT) - 1]$ with $\alpha = (2/\hbar)(\epsilon m^*/N_A)^{1/2}$, is independent of temperature, and

$$I_0 = qp(kT/m^*)^{1/2} \exp(-\alpha\phi_D) \tag{11.22}$$

A plot of $\ln(I_0 T^{-1/2})$ vs $\alpha$ has an activation energy of $q\phi_D$. Experimental data often can be fit by $I_{ti} = C \exp(\beta T) \exp(\alpha\phi)$ where the temperature dependence can be considered to arise from the temperature dependence of $\phi_D$ due to $E_G(T)$.

### Thermally assisted tunneling through interface barrier

This model is similar to the last except that holes tunnel through the barrier into electron-occupied interface states at a hole energy enhanced by thermal excitation. The current is calculated either by integrating over the energy numerically, or by expanding around the maximum.[11] The current is given by $I_{tat} = I_0[\exp(\alpha\phi) - 1]$ with

$$I_0 = I_{00} \exp[-\alpha(\phi_D + E_{fp})] \tag{11.23}$$

with $E_{fp} = E_F - E_v$,

$$\alpha = q/[E_{00} \coth(E_{00}/kT)] \tag{11.24}$$

(note that $\alpha = q/E_{00}$ in the previous case of tunneling without thermal assistance)

$$E_{00} = (q\hbar/2)(N_A/\epsilon m_h{}^*)^{1/2} \tag{11.25}$$

$$I_{00} = (\{4\pi q m^*(kT)^2 \pi^{1/2} E_{00}^{1/2}[q(\phi_D - \phi)]^{1/2}\}/\{h^3 kT \cosh(E_{00}/kT)$$
$$\times [\coth(E_{00}/kT)]^{1/2}\}) \exp[-E_{fp}(1/kT - 1/E_0)] \tag{11.26}$$

with $E_0 = E_{00} \coth(E_{00}/kT)$.

This model results in an $\alpha$ that is weakly temperature dependent. A plot of $\ln\{I_0 \cosh(E_{00}/kT)[\coth(E_{00}/kT)]^{1/2}/T\}$ vs $\alpha$ has an activation energy of $q\phi_D + E_{fp}$.

An example of a heterojunction that appears to be described by this model is n-CdS/p-Zn$_{0.3}$Cd$_{0.7}$Te.[12] Figure 11.10 shows the temperature dependence of $\alpha$ for several different $N_A$ assumed in the thermally assisted tunneling model, indicating a reasonable fit for $N_A = 1.5 \times 10^{19}$ cm$^{-3}$, a density several orders of magnitude larger than the density of acceptors in the bulk ZnCdTe. Figure 11.11 shows the corresponding dependence of the suitably corrected $I_0$ for the thermally-assisted tunneling model as a function of $\alpha$. This model has also been followed in describing the effects seen in ZnO/CdTe and ZnO/InP junctions.[13,14] Tunneling processes appear to dominate

Figure 11.10. Temperature dependence of $1/\alpha$ for an n-CdS/p-ZnCdTe heterojunction, compared to theoretical curves for different values of effective acceptor density $N_A$ for a thermally assisted tunneling model.[12]

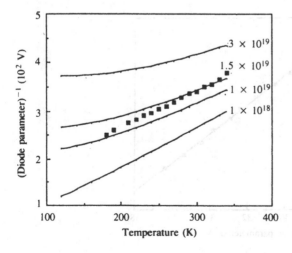

junction transport in real heterojunction structures, particularly in new systems and new materials.

## 11.5    Photovoltaic materials

Photovoltaic effects themselves are common occurrences and can be found wherever there is an internal field in a semiconductor. In order to be practically useful as a solar cell material, however, a semiconductor must have a small enough bandgap to absorb an appreciable portion of the solar spectrum in order to produce a large short-circuit current, and a large enough bandgap to provide a reasonably small value of $I_0$ in the junction current that limits the open-circuit voltage. These considerations limit materials for photovoltaic applications to those with bandgaps in the 1.0–2.0 eV range. Of course, for practical use, solar cells must also be competitively priced, stable under long-term operation, and environmentally benign.

Even with these limitations it might be expected that many materials would be of interest for this kind of application. As a matter of fact only a relatively small number perform efficiently enough to show promise. In single crystal form, for example, an efficiency greater than 10% has been reported only in Si, GaAs, InP, CdTe and CuInSe$_2$. In thin film form, only a-Si:H, CdTe and CuInSe$_2$ (and a few related solid solutions) have been shown to be capable of acting as light absorbing materials to produce photovoltaic cells with efficiency

Figure 11.11. Variation of appropriately corrected $J_0$ for a thermally assisted tunneling model as a function of $\alpha$ for an n-CdS/p-ZnCdTe heterojunction.[12]

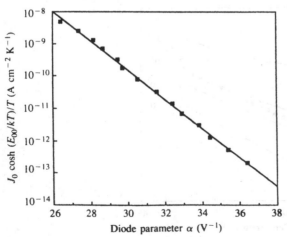

greater than 10%. Leading contenders for use as large bandgap, transparent, high conductivity, n-type window layers, include CdS, ZnCdS, ZnO, $In_2O_3$ and $SnO_2$.

Recent efforts to increase the efficiency of photovoltaic junctions have explored the possibility of developing multijunction cells in which two (or more) different cells are used together to absorb the light more efficiently. The most ideal situation would be to use a large number of such cells in a multijunction such that each cell could effectively absorb light within a narrow range of its band gap and no other. Examples of tested multijunction cells with two components and the efficiencies achieved are: GaAs/Si (31%)[15], GaAs/CuInSe$_2$ (21.3%),[16,17] AlGaAs/GaAs (24–28%),[18-20] a-Si:H/CuInSe$_2$ (15.6%),[21] a-Si:H/a-SiGe:H (13.6%),[22] and GaInP$_2$/GaAs (25%).[23] The structural complexity of even these two-component multijunction cells is illustrated by the cross sectional diagram of Figure 11.12.

Figure 11.12. Solar cell structure for a 23.9% efficient monolithically grown, two-junction solar cell with an $Al_{0.37}Ga_{0.63}As$ top cell and a GaAs bottom cell.[19] Under 1 sun illumination, $\phi_{oc} = 2.41$ V, $J_{sc} = 14.9$ mA cm$^{-2}$, and fill factor = 0.84.

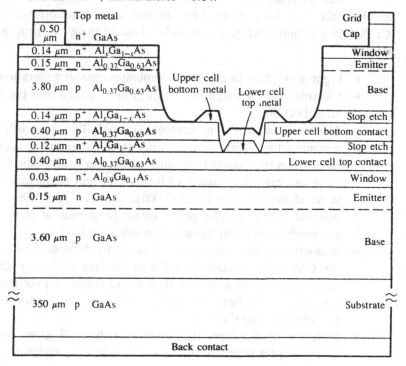

## 11.6    Cu$_x$S/CdS heterojunctions: theater for photoelectronic effects

Many of the photoelectronic properties of semiconductors discussed in this book elsewhere have been observed in research carried out on Cu$_x$S/CdS heterojunctions. The first all thin-film photovoltaic cell, based on a p-Cu$_x$S/n-CdS heterojunction, came close to meeting the minimum efficiency conditions for practical utilization. It can be prepared by a simple dipping process, by placing a crystal or film of CdS into a warm solution containing Cu$^+$ ions and allowing a Cu for Cd replacement reaction to occur. But it exhibited a variety of degradation modes under operation that resisted solution so strongly that ultimately efforts to remedy them have been abandoned. Still, for our present purposes, which are to understand and illustrate photoelectronic effects in semiconductors and not to limit ourselves to commercially successful photovoltaic cells, there are so many examples of photoelectronic effects observable in the Cu$_x$S/CdS system that it is a subject worthwhile discussing.

### Overview of effects

Let us begin with a brief summary of some of the various mechanisms that have been encountered and demonstrated with Cu$_x$S/CdS junctions, before we consider some of them in a little more detail.

* Light absorption in the Cu$_x$S dominates current generation.
* Electrons photoexcited in the Cu$_x$S are injected into the CdS for collection.
* The forward junction current is controlled primarily by recombination and/or tunneling through interface states, rather than by transport into the conduction band of the Cu$_x$S.
* Cu$_x$S can exist in a variety of forms with only small variations in $x$: chalcocite ($x = 1.995-2.000$), djurleite ($x = 1.96$), and digenite ($x = 1.8$). The photovoltaic properties of chalcocite are much superior to those of the other forms.
* Reduction in the value of $x$ can occur by diffusion of Cu into the CdS or by oxidation of Cu at the free surface of Cu$_x$S. Any reduction in $x$ results in a degradation of photovoltaic properties. Any heat treatment of the cell causes appreciable Cu diffusion into the CdS.
* Diffusion of Cu from the Cu$_x$S into the CdS gives rise to deep acceptor states in the CdS that widen the depletion layer

width, and whose occupancy and charge can be affected by photoexcitation.

* Response to low-energy photons is increased by simultaneous photoexcitation with high-energy photons because of the photoexcited change in charge in the depletion region and the effects of this on current collection. This gives rise to an *enhanced* state when the depletion layer is narrow, and a *quenched* state when the depletion layer is wide.

* Dark and light forward bias *I* vs *V* curves often cross because of a change in junction current mechanism under illumination. This effect is strongly accentuated if the cell has been subjected to a heat treatment.

* An additional degradation process occurs through a reversible photochemical defect reaction, caused by photoexcitation after Cu diffusion into the CdS has occurred. This gives rise to a *degraded* state when lifetime-killing defects have been formed by this process, and a *restored* state when these defects have been thermally annealed away.

A general energy band diagram for the Cu$_x$S/CdS heterojunction is given in Figure 11.13. The diagram assumes that the Fermi level lies at the valence band edge in degenerate p-type Cu$_x$S and 0.1 eV below the conduction band edge in n-type CdS. Conduction band and valence band discontinuities at the interface are downward, and no energy spikes are predicted.

The Cu$_x$S/CdS junctions described in this section consist of a film of Cu$_x$S formed on the surface of a single crystal of CdS.

*Junction capacitance*

If a Cu$_x$S/CdS cell is subjected to a mild heat treatment for 1 min at 250 °C in air, a considerable increase in the depletion layer width is measured (e.g., from 0.20 to 0.63 $\mu$m at zero bias).

*Spectral response*

The spectral response of the short-circuit current is strikingly changed by this heat treatment, as shown in Figure 11.14. Before heat treatment the spectral response extends to about 1.0 $\mu$m at long wavelengths; after heat treatment the response is decreased over the whole spectral range and the long-wavelength cutoff is at 0.7 $\mu$m. If the junction is illuminated with 500 nm light while the spectral response is being measured, the long-wavelength response is restored with no effect at short wavelengths.

In the measurement of the spectral response for the heat treated junction, long time effects were observed. For wavelengths greater than 0.83 μm, the photocurrent decreased slowly from its initial value to a steady state value, whereas for shorter wavelengths, the photo-current increased slowly from its initial value to a steady state value.

### Optical quenching

The effects of secondary infrared radiation on the short-circuit current are shown in Figure 11.15. Strong quenching of the primary photocurrent is observed in two bands corresponding to transitions at 0.8 and 1.1 eV, corresponding approximately to optical quenching

Figure 11.13. Approximate energy band diagram of a p-Cu$_x$S/n-CdS heterojunction.[1]

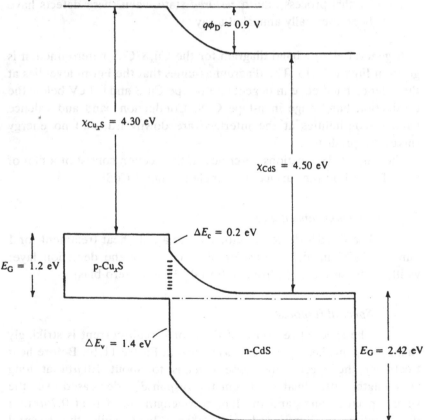

transitions in high-sensitivity CdS crystals, as shown in Figures 5.8 and 5.18(*b*).

### Persistence of enhancement

The long-wavelength enhancement of photocurrent shown in Figure 11.14 as a result of illumination with short-wavelength light persists for long times after irradiation with short wavelength is terminated. The temperature dependence of this persistent current was measured by plotting the initial dark decay rate against reciprocal temperature; the data show an activation energy of 0.95 eV and a hole capture cross section for the imperfections involved of $2.5 \times 10^{-15}$ cm$^2$.

### Enhancement/quenching model

The phenomena described thus far can be understood in terms of the energy band diagram of Figure 11.16. Heat treatment causes Cu to diffuse into the CdS, giving rise to an increase in negative charge, a

Figure 11.14. Spectral response of the short-circuit current before and after heat treatment of a Cu$_x$S/CdS cell for 1 minute at 250 °C, and the spectral response after heat treatment measured with a 500 nm secondary illumination.[24] The intensity of both primary and secondary light sources is 300 $\mu$W cm$^{-2}$.

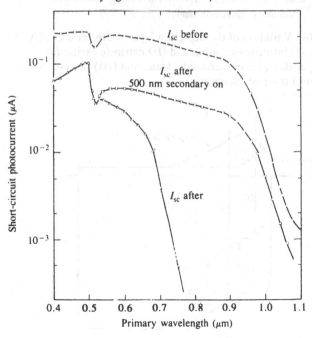

Figure 11.15. Enhancement and quenching of photocurrent by secondary light for a heat-treated $Cu_xS/CdS$ junction.[24] Arrows indicate where the onset of optical quenching of photoconductivity occurs in sensitive CdS crystals as shown in Figure 5.8. Both the 0.655 $\mu$m primary and the secondary intensity were 300 $\mu$W cm$^{-2}$.

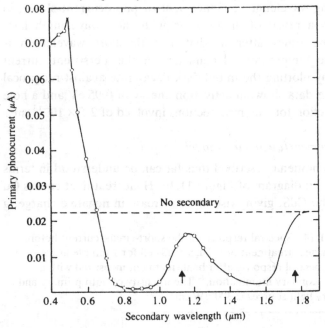

Figure 11.16. Variation of the depletion layer width for a $Cu_xS/CdS$ junction by (I) intrinsic excitation and (II) extrinsic excitation. Quenching of short-circuit current by transition (III) represents either an optical or a thermal excitation.

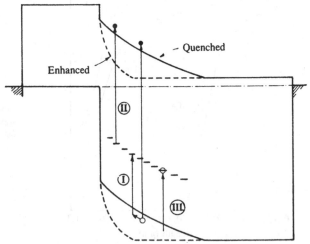

decrease in junction capacitance, a widening of the depletion layer, and the behavior associated with a *quenched* state of the junction. As a consequence, the short-circuit current decreases due to recombination loss through interface states. Intrinsic photoexcitation creates holes that are captured by these acceptors, or suitably energetic extrinsic photoexcitation excites directly from these acceptors to the conduction band, thus decreasing the negative charge in the depletion layer, increasing the junction capacitance, making the depletion layer narrower, and increasing the short-circuit current. This is the *enhanced* state of the junction. Optical quenching removes holes from the acceptors in the depletion region, causing the enhanced state to change to the quenched state. Observing the dark decay of persistence gives a thermal activation energy corresponding to the height of these acceptor levels above the valence band.

### Photocapacitance effects

Even in a $Cu_xS/CdS$ junction that has not been heat treated after preparation, illumination of the junction with bandgap light at low temperatures produces a significant increase in the junction capacitance, which persists after the removal of the photoexcitation. We call 'photocapacitance,' $\Delta C$, the difference between the capacitance after photoexcitation and the dark capacitance. A typical excitation spectrum for this photocapacitance is shown in Figure 11.17. The results show that photocapacitance can be excited by intrinsic and extrinsic (0.5–1.0 $\mu$m) photoexcitation, following the description given in Figure 11.16; as well as by light with wavelengths out to 1.4 $\mu$m. The relatively small effect in this latter wavelength range is probably associated with photoexcitation plus tunneling transitions. The magnitude of the measured photocapacitance increases upon cooling until at about 130 K the photocapacitance is about 60% of the dark capacitance.

The photocapacitance can be reduced by optical quenching. Typical quenching spectra at two temperatures are given in Figure 11.18. This experiment was performed by first establishing $\Delta C$, then shining on monochromatic light for a fixed time and recording the decrease in $\Delta C$. For the measurements at 250 K, two distinct quenching bands are resolved, centered at 0.9 and 1.35 $\mu$m, with clear similarity to the optical quenching spectra for photocurrent in CdS and CdS:Cu, given in Figures 5.8, 5.18(b), and 11.15.

The photocapacitance can also be reduced by applying bias to the junction. A variety of bias effects with or without concurrent optical quenching of photocapacitance are given in Figure 11.19. Reverse bias in the dark or with 1.35 $\mu$m light decreases $\Delta C$ only slightly. Reverse bias with 0.85 $\mu$m light causes a significant reduction in the portion of the $\Delta C$ remaining after optical quenching by 0.85 $\mu$m light alone, probably by preventing retrapping of holes optically freed from acceptors in the optical quenching process. Forward bias in the dark or with 1.35 $\mu$m light causes a large decrease in $\Delta C$. This decrease in $\Delta C$ is probably caused by tunneling of electrons from the CdS conduction band to interface states, followed by recombination of some of these electrons with holes trapped in the deep levels nearest to the interface. Forward bias with 0.85 $\mu$m light (after the full extent of optical quenching by this wavelength) causes only a small additional decrease in $\Delta C$.

Figure 11.17. Photocapacitance excitation spectrum for an unheated Cu$_x$S/CdS junction at 105 K.[25]

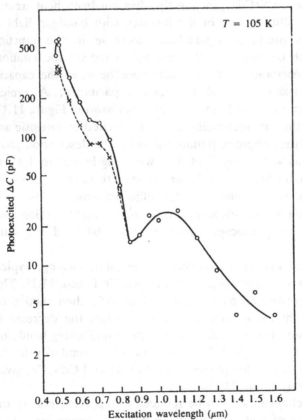

The presence of the photocapacitance, indicating a narrower depletion layer, is associated with an increase in the forward bias current, as
is suggested by the energy band of Figure 11.16 with forward currents
flowing by tunneling to interface states.

It is possible to make a close correlation between the magnitude of
the photocapacitance and the magnitude and spectral response of
short-circuit current in both unheated and heat treated $Cu_xS/CdS$
junctions. Data for the unheated junction described in Figures
11.17–11.19 are given in Figure 11.20, while comparable data for a
heat-treated junction are given in Figure 11.21. The spectral response
of short-circuit current at 105 K is shown in Figures 11.20(*a*) and
11.21(*a*), with and without $\Delta C$. In the state without $\Delta C$, the
short-circuit current in the unheated junction is about four times
smaller than the current at 300 K, while for the heat-treated junction
the current is about 200 times smaller. In both cases the creation of the

Figure 11.18. Optical quenching spectra for photocapacitance for an
unheated $Cu_xS/CdS$ junction for white light, and 500 nm, primary
photoexcitation at 110 K, and for 520 nm primary photoexcitation at
250 K.[25]

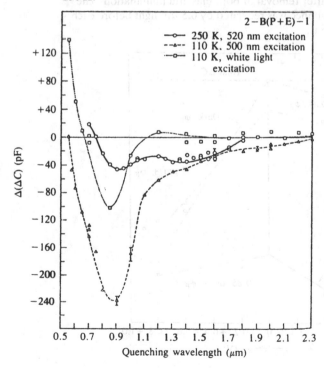

$\Delta C$ state increases the short-circuit current to close to the current at 300 K. The measured short-circuit current as a function of $\Delta C$ is plotted in Figures 11.20($b$) and 11.21($b$).

### Thermally restorable optical degradation (TROD)

Another phenomenon associated with the properties of the depletion layer in heat treated $Cu_xS/CdS$ junctions is a large and rapid degradation process introduced by optical excitation after heat treatment, which can be reversed by thermal annealing at about 100 °C in the dark.

The effect appears to be closely related to a decrease in electron lifetime on optical illumination of Cu-doped single crystal CdS.[26]

Figure 11.19. Effects of applied forward or reverse bias and quenching light on the photocapacitance in an unheated $Cu_xS/CdS$ junction at 110 K.[25] Three cases are included: ($i$) forward or reverse bias applied in the dark; ($ii$) 1.35 $\mu$m quenching light applied for 300 s with no bias, and left on during application of forward or reverse bias; ($iii$) the same as ($ii$) for 0.85 $\mu$m quenching light. In all cases the bias was applied for 100 s and the final value of capacitance was measured after removal of both bias and illumination. The photocapacitance was reexcited by 0.5 $\mu$m light before each measurement.

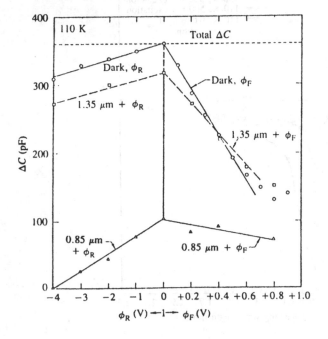

Although the exact mechanism of the effect cannot be said to be known, it is apparently closely related to the photochemical reactions described in Section 8.7. The combination of photoexcitation to change the electronic state and thermal energy sufficient to cause ionic motion produces a new center with a large capture cross section for free electrons; subsequent thermal annealing dissociates this center into its original constituents.

The most direct observation of the effect is the decrease in short-circuit current for a heated Cu$_x$S/CdS junction, or the photocurrent for a CdS:Cu crystal, with time under constant illumination. The total variation of current vs time is shown in Figure 11.22(a); the current rapidly rises to a maximum when the photoexcitation is turned on, reaches a maximum, and then decays slowly as the light-induced degradation process dominates. The time decay portion of the curve is

Figure 11.20. Effect of photocapacitance $\Delta C$ on (a) the spectral response of short-circuit current, measured with and without $\Delta C$ at 105 K, compared to the response at 300 K, and (b) the short-circuit current measured at 0.7 $\mu$m at 105 K, for two different light intensities, for an unheated Cu$_x$S/CdS junction.[25]

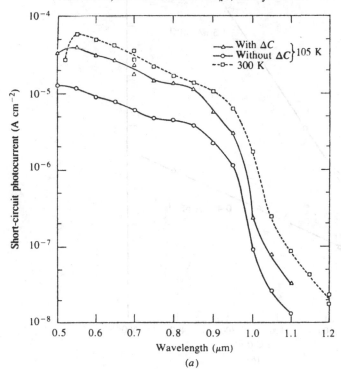

(a)

shown in Figure 11.22(*b*) for three junctions and one single crystal. Optical degradation leads irreversibly to a *degraded* state that is stable until the sample is thermally annealed to its *restored* state. These degraded and restored states are independent of the enhanced and quenched states of the junction described above, making possible four distinct states: restored-enhanced (RE), degraded-enhanced (DE), restored-quenched (RQ), and degraded-quenched (DQ).

The formation of the degraded state requires sufficient thermal energy as well as photoexcitation, but beyond a certain temperature thermal effects anneal the state away. Plots of degradation as a function of temperature are given in Figure 11.23(*a*), and plots of thermal restoration as a function of temperature are given in Figure

Figure 11.20. (*cont.*)

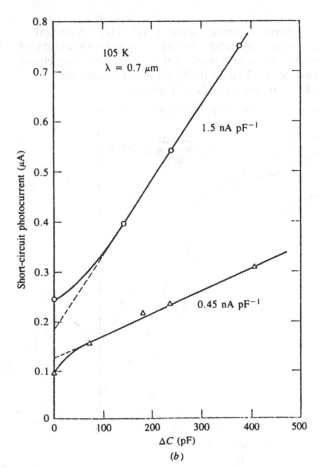

105 K
λ = 0.7 μm

1.5 nA pF⁻¹

0.45 nA pF⁻¹

Short-circuit photocurrent (μA)

ΔC (pF)

(*b*)

11.23(*b*). Optical degradation of the junction begins at about 200 K and is virtually complete at about 230 K for the conditions chosen (25 s exposure to 300 $\mu$W cm$^{-2}$ of 0.535 $\mu$m light). For the same conditions optical degradation of the single crystal starts at 250 K and is complete at about 330 K. The process has an activation energy of about 0.4 eV in both junction and single crystal. Thermal restoration occurs at temperatures above about 350 K and is also very similar for the

Figure 11.21. Effect of photocapacitance $\Delta C$ on (*a*) the spectral response of short-circuit current, measured with and withour $\Delta C$ at 105 K, compared to the response at 300 K, and (*b*) the short-circuit current measured at 0.7 $\mu$m at 105 K, for the same Cu$_x$S/CdS junction as described in Figures 11.17–11.20 after heat treatment at 200 °C for 2 min.[25]

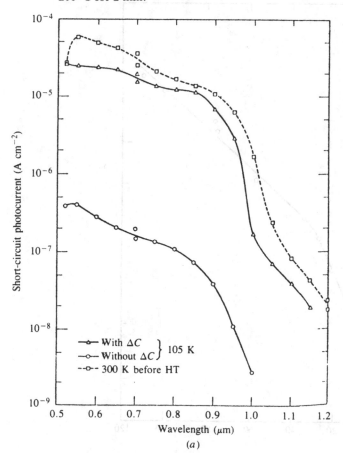

(*a*)

junction and the single crystal, as shown in Figure 11.23(*b*); the activation energy for thermal restoration is of the order of 1.6 eV.

In spite of changes in the magnitude of the photocurrent by as much as a factor of $10^3$ between DQ and RE states for the $Cu_xS/CdS$ junction, the relative shape of the spectral response curves is approximately unchanged. In all states the short-circuit current and the forward junction current are related so that they increase or decrease together; this is in agreement with a model in which the forward current flows by tunneling and recombination through interface states to the $Cu_xS$ valence band, and collection of the short-circuit current is

Figure 11.21. (*cont.*)

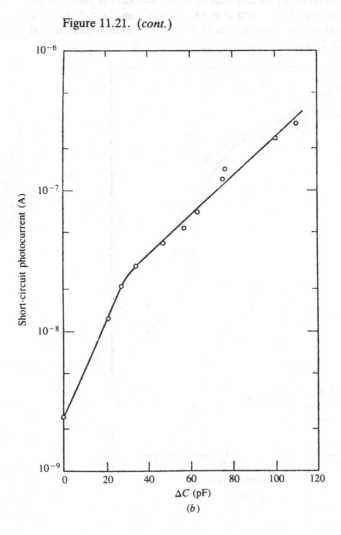

(*b*)

Figure 11.22. (*a*) Schematic Thermally Restorable Optical
Degradation (TROD) curve showing the variation of photocurrent
with time after the beginning of photoexcitation. (*b*) Short-circuit
current of two heated $Cu_xS/CdS$ junctions, and photocurrent of a
single crystal of CdS:Cu as a function of time during optical
degradation. Time scales for the four curves from top to bottom are
200, 4000, 4000, and 100 s div$^{-1}$, respectively. The upper three curves
are for 300 K. Fully degraded values of current are indicated on the
right-hand side for two of these curves.[27]

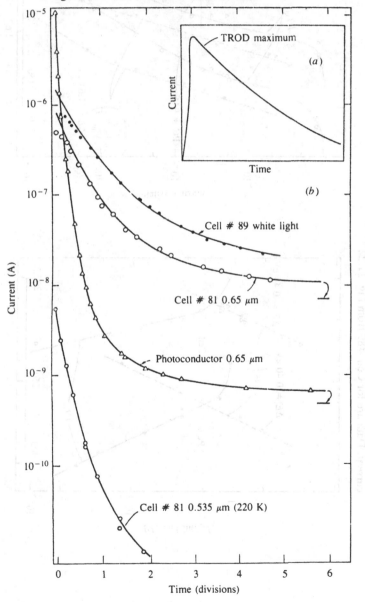

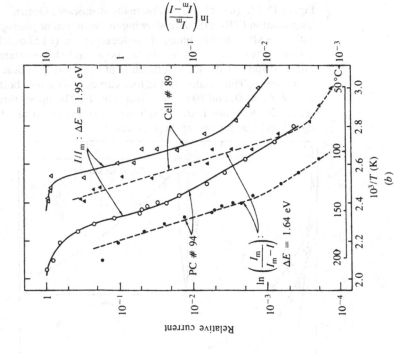

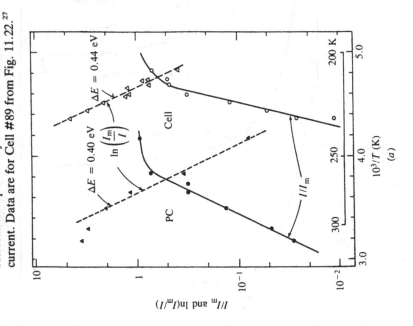

Figure 11.23. (a) Optical degradation vs temperature for a $Cu_xS/CdS$ junction and for a photoconductor, relative to the fully restored state. The degree of restoration is probed with a 0.65 $\mu$m light at 150 K. $I$ is the probe current and $I_m$ is the fully restored value. Data are for Cell #81 from Figure 11.22. (b) Thermal restoration vs temperature of restoration for a $Cu_xS/CdS$ junction and for a photoconductor relative to the fully restored state. The degree of restoration is probed by the TROD curve maximum $I$ at 300 K by using white light. $I_m$ is the fully restored value of current. Data are for Cell #89 from Fig. 11.22.[27]

increased when the field at the interface is large enough to prevent recombination of photoexcited carriers through interface states. The qualitative model is given in Figure 11.24. Enhancement and quenching effects are controlled by the field at the interface; degradation and restoration effects are controlled by the properties of the high-resistivity layer adjacent to the junction interface.

Finally the entire reversible TROD cycle is pictured in Figure 11.25 from 150 to 500 K. The cycle starts at room temperature with the junction in the DQ state. As the temperature is decreased, the junction goes to the DE state due to a decrease in thermal quenching; little hysteresis occurs upon returning to 300 K. From 325 to 375 K, the current drops to unmeasurably small values, as the state changes from DE to DQ, and then increases as the temperature is raised and thermal restoration produces the RQ state. As the temperature is decreased from 475 K, the enhancement/quenching ratio increases and the junction enters the RE state with also a small effect due to optical

Figure 11.24. Simplified energy band diagram of the Cu$_x$S/CdS interface to illustrate the control of short-circuit current and forward bias current under the effects of enhancement and quenching.

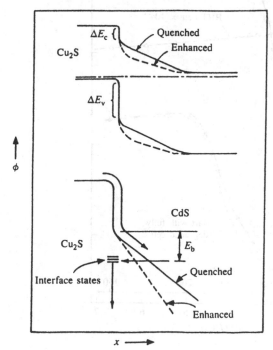

degradation. A qualitatively identical cycle is exhibited by the photo-current in a single crystal.

Also shown is the current for a 'before heat treatment' (BHT) junction, multiplied by $10^{-3}$ to fit on the plot conveniently, and data for another junction obtained after 1 min heat treatment at 160 °C,

Figure 11.25. A complete thermal cycle of short-circuit current excited by white light for Cell #89 from Figure 11.22, after long heat treatment and an intensity of 1400 $\mu$W cm$^{-2}$, together with an indication of the magnitude of the TROD maximum representing the fully restored state.[27] Also shown is the current for a 'before heat treatment' (BHT) junction, multiplied by $10^{-3}$ to fit on the plot conveniently, and data for another junction obtained after 1 min heat treatment at 160 °C, multiplied by $6.1 \times 10^{-3}$ to normalize to the BHT data.

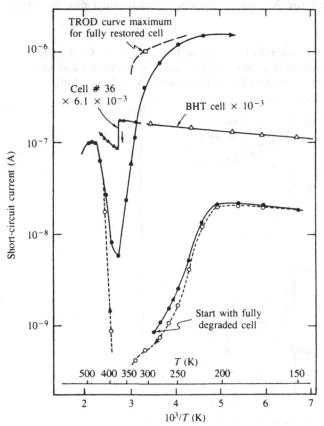

multiplied by $6.1 \times 10^{-3}$ to normalize to the BHT data. These data show an abrupt step in current due to a phase transformation at 78–82 °C (orthorhombic to tetragonal transformation in djurleite Cu$_{1.96}$S), but there is no rapid variation in current at temperatures on either side of the step.

# 12

## Quantum wells and superlattices

In order to be useful for photoelectronic applications, a semiconductor must be prepared with the appropriate purity and crystallinity. Such a preparation treatment may be considered the first technological step in designing semiconductor materials. In this book we have described a number of different photoelectronic effects that depend on the properties of impurities and crystal structure in bulk and thin-film semiconductors.

A second step developed with the recognition that striking new potentialities become possible if semiconductors are grown or treated in such a way that junctions are formed – leading then to the development of a variety of Schottky junctions, homojunctions, heterojunctions, buried homojunctions, and p–i–n, MIS, and SIS junctions. Some of the properties of such junctions in photodetectors and photovoltaic applications have also been discussed in this book.

Finally we come to some of the most recent steps in shaping the properties of materials to meet specific design criteria: direct 'bandgap engineering,' made possible by the development of highly controllable deposition systems for thin films, such as molecular beam epitaxy (MBE) and organometallic chemical vapor deposition (OMCVD). These techniques have made it possible to deposit multilayer heterojunctions with atomically abrupt interfaces, and controlled composition and doping in individual layers that are only a few tens of nanometers thick. These layers are so thin that the energy levels in them in their small dimension direction show the quantum effects predicted for an electron in a very narrow potential well. Their properties depend directly on the wave nature of the electron. They include heterojunction and doped superlattices, modulation-doped superlattices, and strained layer and variable-gap superlattices.[1-4]

Through the use of band-gap grading, one can make effectively continuous changes in a particular energy band diagram. The transport properties of electrons and holes can be independently tuned to meet specific needs.

A quantum well is formed when a thin layer of a small bandgap material is sandwiched between two thin layers of a larger bandgap material, forming a heterojunction structure. The specific values of the discrete levels in the quantum well are determined by the thickness and the depth of the well, which is the conduction band discontinuity $\Delta E_c$ for electrons, and valence band discontinuity, $\Delta E_v$ for holes.

If many quantum wells are grown on top of one another, and the barriers are made so thin (<5 nm) that tunneling between them is significant, the result is a superlattice with properties controlled by the artificial periodicity of the multilayer structure. The properties of superlattices can be varied by choices of the materials involved, the thickness of the layers used, and the spacing between layers. It becomes possible to fine tune a whole range of properties by variations of these compositional and structural parameters.

Any attempt at a comprehensive discussion of such quantum wells and multilayer structures could be encompassed only by an entire book or, more likely, several books.[1,2] In this chapter we consider the possibilities of only a subset of such structures, limiting ourselves to those that are directly relevant to photoelectronic applications.

## 12.1 Electron in a one-dimensional quantum well

The simplest model is for an electron in a one-dimensional quantum well defined by $V = 0$ for $0 \leq x \leq L$ and $V = \infty$ for $x < 0$ or $x > L$. The solution of the Schroedinger equation is

$$\psi = A \exp(ikx) + B \exp(-ikx) \tag{12.1}$$

with

$$E = \hbar^2 k^2 / 2m \tag{12.2}$$

When subject to the boundary conditions that $\psi = 0$ for $x = 0$ and $x = L$,

$$k = n\pi / L \tag{12.3}$$

where $n$ is an integer, and

$$E_n = n^2 h^2 / 8mL^2 \tag{12.4}$$

The allowed energies are a discrete set with spacing depending on $1/L^2$. Substituting numerical values for the constants gives $E_1 = 4 \times 10^{-15}/L^2$ eV. If $L = 1$ cm, then $E_1 \approx 0$, but if $L \leq 40$ Å, $E_1 \geq kT$ at room temperature.

This problem can be generalized to make it more realistic by considering $V = 0$ for $0 \leq x \leq L$ and $V = V_0$ for $x < 0$ or $x > L$. If $x < 0$, the mathematically well-behaved solution is

$$\psi_1 = A_1 \exp(\alpha x) \quad \text{with } \alpha = [2m(V_0 - E)]^{1/2}/\hbar \tag{12.5}$$

Similarly for $x > 0$, the mathematically well-behaved solution is

$$\psi_3 = B_3 \exp(-\alpha x) \tag{12.6}$$

For $0 \leq x \leq L$

$$\psi_2 = A_2 \exp(i\beta x) + B_2 \exp(-i\beta x) \quad \text{with } \beta = (2mE)^{1/2}/\hbar \tag{12.7}$$

After application of the boundary conditions for continuity of the wavefunction at $x = 0$ and $x = L$, allowed energy levels are given by

$$\tan[(L/\hbar)(2mE)^{1/2}] = 2[E(V_0 - E)]^{1/2}/(2E - V_0) \tag{12.8}$$

The number of allowed energy states within the well increases with increasing $L$. For very small $L$ there is finally only one, but always at least one, allowed energy state.

## 12.2    Effective mass filtering photoconductor

Our discussion of Figures 2.5($d$) and 2.8 showed how large photoconductivity gain can be realized in two different structures. Common to both was localization of the hole, so that many electron transits of the material could be made before either recombination with the hole or the hole became delocalized and left the material. In Figure 2.5($d$) the hole was localized at an imperfection in a homogeneous material with ohmic contacts, and in Figure 2.8 the hole was localized in the potential well of an n–p–n junction. The photoconductivity gain was given by the ratio of the electron lifetime to transit time.

An analogous effect can be produced with a suitably designed superlattice structure, when the superlattice is used to produce 'effective mass filtering.' This is the result of effective localization of heavy-mass photoexcited holes in the superlattice wells, while the lighter-mass photoexcited electrons are relatively free to move by phonon-assisted tunneling, or in an extended state known as a miniband, throughout the material.[5] An appropriate energy band diagram for the photoconductor is given in Figure 12.1.

Structures tested were grown by MBE on $\langle 100 \rangle n^+$-InP and consisted of an undoped $Al_{0.48}In_{0.52}As$(35 Å)/$Ga_{0.47}In_{0.53}As$(35 Å) 100 period superlattice sandwiched between two degenerately doped $n^+$ 0.45 $\mu$m thick $Ga_{0.47}In_{0.53}As$ layers. They were responsive to light in the infrared at 1.0–1.6 $\mu$m. Significant gains of the order of 50 were obtained at applied voltages as low as 20 mV, and a gain of $2 \times 10^4$ was measured for 1.4 V. The response time was about $10^{-4}$ s, and the noise was low due to the low-voltage operation.

Because of the small layer thicknesses used, the wells are coupled and the quantum states form minibands. Since the tunneling probability varies inversely exponentially as the effective mass, the electrons have a more than two order of magnitude higher tunneling probability than the heavy holes. In addition, electrons can be easily injected into the superlattice from the $n^+$ contact region. Since carrier mobility in the miniband (in the above specific system, $\mu_n = 0.15 \, cm^2 \, V^{-1} \, s^{-1}$ typical of a phonon-assisted tunneling conduction) depends exponentially on the superlattice barrier thickness, the electron transit time, the photoconductive gain, and the gain-bandwidth product can be artificially tuned over a wide range. This gives a design flexibility not available with standard photoconductors.

## 12.3 Avalanche photodiodes (APDs)

Photodetectors with gain can also be made by involving a process known as avalanche multiplication in a special type of p–n junction. In an avalanche-multiplication photodiode (APD), photoexcited carriers receive sufficient energy from the electric field to raise electrons from the valence band to the conduction band by impact ionization. The energy required to do this is larger than the bandgap,

Figure 12.1. Band diagram of a superlattice photoconductor, showing effective localization of photoexcited holes in quantum wells, while electrons tunnel in a miniband.[3]

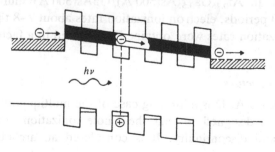

and is known as the ionization threshold energy. The carriers produced by impact ionization then engage in additional impact ionization excitation of carriers, resulting in an avalanche.

One of the disadvantages of a standard APD is that excess noise is generated as a result of fluctuations in the avalanche gain if both electrons and holes are involved in the impact ionization process. A feedback process exists in which any slight fluctuation in the electron multiplication process is fed back and amplified because of the hole-initiated impact ionization. If only one carrier is involved in impact ionization, however, the avalanche proceeds in only one direction, this feedback does not exist, and excess noise is greatly reduced. It is therefore desirable to design an APD with impact ionization of only electrons, or at least one in which the ionization rate of electrons is much larger than that of holes. Crystalline Si fulfils this criterion with an ionization rate for electrons 20 times that for holes, but the spectral region of sensitivity for Si is limited. All III–IV compounds have approximately equal ionization rates for electrons and holes.

By the use of 'bandgap engineering', however, it is possible to design the ionization rates of electrons and holes, rather than simply accepting values given in a particular material. Several types of superlattice structures have been developed with these goals in mind.[1,3]

### Multiquantum wells

A superlattice alternates material with high and low bandgaps and limits ionizing collisions to the low bandgap material, as shown in Figure 12.2. Electrons and holes are both accelerated in the wide bandgap regions but do not receive enough energy to ionize. When it enters the next well, the electron receives enough energy from the conduction band discontinuity $\Delta E_c$ to ionize, but the same does not happen for the hole since the valence band discontinuity $\Delta E_v$ is not large enough. In an $Al_{0.45}Ga_{0.55}As(500 \text{ Å})/GaAs(500 \text{ Å})$ multiquantum APD with 50 periods, electron ionization rates about 7–8 times as large as hole ionization rates were obtained, compared to a factor of 2 in GaAs itself.[6]

### Staircase structures

The staircase APD is a limiting case of the multiquantum well APD, in that it is designed so that the whole ionization energy is gained at the band discontinuity. It is considered an archetype of 'bandgap engineering' at its best,[1,3] and corresponds in many ways to the vacuum tube photomultiplier with high internal amplification and negligible avalanche noise. Figure 12.3(*a*) shows the band diagram of a graded-gap multilayer structure (assumed intrinsic) without applied bias, and Figure 12.3(*b*) shows the corresponding situation with an applied bias. Each step is linearly graded in composition from a low $(E_{G1})$ to a high $(E_{G2})$ bandgap, with an abrupt step back to low bandgap material. The materials selected are chosen so that the conduction band discontinuity $\Delta E_c$ is larger than the electron ionization energy in the small bandgap material. In a practical device, an ungraded layer should be left immediately after the step with a thickness of a few ionization mean free paths (50–100 Å), so that most electrons ionize near the step. In an alternate design one might choose a bandgap discontinuity $\Delta E_c$ smaller than the ionization energy, and provide the additional required energy by leaving an ungraded region before each step.

The electron ionization occurs at each step of the staircase, not in the graded-gap regions where the electric field is too small. Holes are multiplied by the electron impact ionization, but they do not themselves cause ionization since the bandgap discontinuities are of the wrong sign to cause ionization, and ionization does not occur in the valence band because the electric field is too small. Because of the absence of hole ionization, the excess noise is reduced to a lower value than in any standard APD. Research on reducing this structure to practice continues with HgCdTe or III–V alloys as likely choices.

Figure 12.2. Band diagram for a multiquantum well APD, limiting ionizing collisions to the low bandgap regions only.[3]

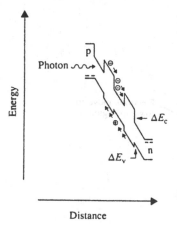

Figure 12.3. (*a*) Band diagram for a staircase APD with each step having a graded bandgap, for no applied bias. (*b*) The staircase APD with applied bias.[1]

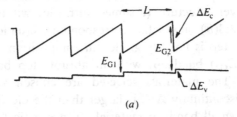

(*a*)

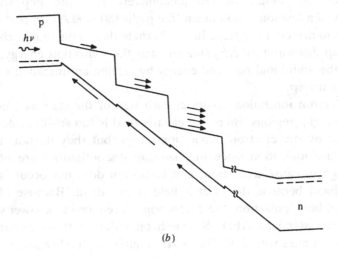

(*b*)

Figure 12.4. The three-dimensional band diagram for a channeling APD.[3]

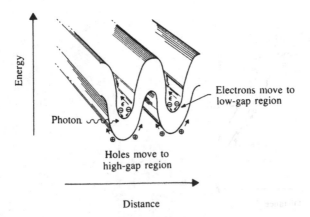

### Channeling structures

Another multilayer APD device consists of alternating p- and n-type layers with different bandgap, with lateral p- and n-type contacts, as shown in Figure 12.4.[3] The band boundaries are like two gutters separated by a space defined by the bandgap. Channels formed by the p–n junction run parallel to the layers. A periodic electric field perpendicular to the layers results from the alternating npnp layers and an external field is applied parallel to the layers. The transverse electric field collects electrons in the low-gap n-type regions, where they are channeled by the parallel electric field, so that electrons impact ionize in the low-gap n-type regions. In addition, the transverse field also causes holes created by electrons ionizing in the n-type regions to transfer into the wide-gap p-type regions before ionizing. There holes cannot impact ionize because the bandgap is too high. Therefore multiplication occurs primarily by electrons only and the noise is small.

### Impact ionization across the band discontinuity

Another type of proposed superlattice APD is one in which impact ionization occurs across the band discontinuity, as shown in Figure 12.5. High-energy electrons in the barrier layer interact with carriers confined in n-type doped wells and impact ionize them out across the band-edge discontinuity. The process has similarities with impact ionization of electrons localized in deep imperfection states in a semiconductor. Since only electrons are produced by the impact

Figure 12.5. Impact ionization across the band-edge discontinuity.[1] (a) For quantum wells doped *n*-type. (b) For undoped wells, in which the carriers originate from thermal generation processes through midgap centers.

Hot electron

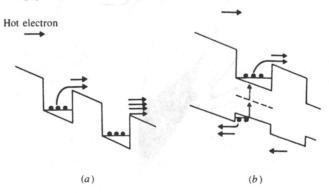

(a)                    (b)

ionization process, there is small excess noise. Of course suitable contacts must be provided to the wells in this case to replenish the electrons.

As indicated in Figure 12.5(*b*), the effect can also be realized with undoped wells; in this case the thermal generation of electrons and holes in the wells can lead to relatively large electron and hole densities dynamically stored in the wells if the band discontinuities are large compared to the average carrier energies. Such a situation has been realized in the high-field region of quantum well p–i–n structures made with an $Al_{0.48}In_{0.52}As/Ga_{0.47}In_{0.53}As$ superlattice in the i-region with well thicknesses between 100 and 500 Å.[7] Results indicate that holes ionize at a considerably higher rate than electrons in this structure.

If graded bandgap structures are included, other variations are possible. For example, Figure 12.6 shows a superlattice in which the interface of the wells has been appropriately graded to eliminate the

Figure 12.6. (*a*) Band structure of a multiple graded-well photomultiplier with the graded regions shaded, showing multiplication of holes.[1] (*b*) Mechanism of hole multiplication by impact ionization across the band-edge discontinuity.

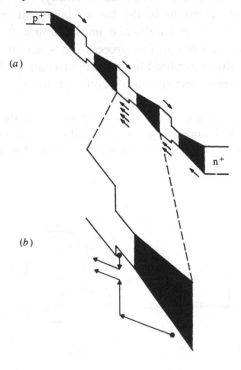

storage of electrons, while preserving that of holes. Subsequent impact ionization is almost completely by holes, again producing a low noise situation. A corresponding structure has been grown by MBE with a three-period superlattice i-region of a p–i–n photodiode:[8] 501 Å $Al_{0.48}In_{0.52}As$ barriers, 292 Å $Ga_{0.47}In_{0.53}As$ wells, and 1022 Å AlInGaAs graded regions. The hole ionization rate, produced by avalanche ionization, was measured to be 50 times the electron ionization rate, controlled by ionization across the bandgap.

## 12.4 Quantum well infrared photodetectors

In a quantum well superlattice structure there are also extended states above the barriers. Quantum well infrared photo-detectors (QWIPs) have been obtained by photoexciting electrons from a single bound state in the wells to the extended states and hence with transport to the contacts, as shown in Figure 12.7.

Typical devices of this type are (1) 50 periods of 38 Å GaAs quantum wells doped to $n = 2 \times 10^{18}$ cm$^{-3}$, and 280 Å of $Al_{0.22}Ga_{0.78}As$ barriers – with sensitivity in the 8–12 $\mu$m range;[9,10] (2) 50 periods of 30 Å $In_{0.53}Ga_{0.47}As$ quantum wells doped to $n = 2 \times 10^{18}$ cm$^{-3}$, and 300 Å of undoped $In_{0.52}Al_{0.48}As$ as barriers – with sensitivity in the 3–5 $\mu$m range.[11]

It is possible with these structures to achieve a photoconductive gain greater than unity.[12] QWIPs with twenty quantum well periods of 40 Å GaAs doped to $n = 2 \times 10^{18}$ cm$^{-3}$ and 500 Å barriers of undoped $Al_{0.3}Ga_{0.7}As$, were compared with equivalent QWIPs with only two quantum well periods. These detectors had n$^+$-GaAs ohmic contacts at the ends of the quantum well structure, outside of the $Al_xGa_{1-x}As$ barriers. Both detectors had a peak responsivity at 7.1 $\mu$m at 25 K, but

Figure 12.7. QWIP in which electrons are photoexcited from the wells to extended states above the barriers.[4]

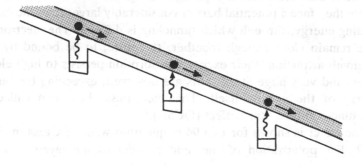

the gain (8.1) for the two quantum well structure was an order of magnitude larger than that (0.59) for the twenty quantum well structure. These results show that the photocarrier lifetime is not limited by the transit time, and that the gain is inversely proportional to the quantum well width. The transport mechanism is therefore analogous to that in conventional photoconductors; the photoexcited electrons can circulate via reinjection from the emitter contact, until the 'hole' is refilled by electron capture into the quantum well.

## 12.5    Quantum confined Stark effect devices

A number of applications of quantum well structures are based upon the effects of an electrostatic field applied to the electrons in these structures. Of particular interest to our discussion here is the existence of exciton absorption even at room temperature due to the quantum confinement of the electron and hole in a thickness much less than the normal exciton diameter, and the Stark effects of an electrostatic field applied to such an exciton.

Figure 12.8 shows schematically the effects of an electric potential on the excitonic bound states. If the electric field is applied parallel to the layers, the situation is like that in bulk material, and the applied field induces Stark effects in the exciton ground state. At lower electric fields one observes a broadening of the exciton resonance absorption due to a decrease in exciton lifetime because of field ionization, but at readily achievable fields (of the order of $10^4 \, \text{V cm}^{-1}$) this effect can become so large that the resonances are no longer resolvable. The Stark shifts involved are usually so small that they are not detected.

If the electric field is applied perpendicular to the layers, however, effects result that are unique to the quantum well structures, provided that the layers are not thick enough to result in destruction of the resonance by field separation of electrons and holes. For layers with thickness less than the three-dimensional Bohr radius of the exciton, the electrons and holes are pushed up against the sides of the well where they face a potential barrier considerably larger than the exciton binding energy, through which tunneling is difficult. The electron and hole remain close enough together, therefore, to be bound by their Coulomb attraction, their excitonic absorption persists to high electric fields, and very large Stark shifts are observed, exceeding the binding energy of the exciton itself. This mechanism has been called the quantum-confined Stark effect (QCSE).[13]

The effect is shown for two 94 Å quantum wells in GaAs in Figure 12.9. For polarization of the light parallel to the layers, excitonic absorption is observed corresponding to both light and heavy holes, but for polarization perpendicular to the layers only the excitons involving light holes are observed. For an electric field of $2.2 \times 10^5$ V cm$^{-1}$, shifts in exciton absorption by up to 40 meV are observed, ten times the bulk exciton binding energy.

For device applications these layers are usually incorporated as the i-layer of a p–i–n junction diode, in order to enable the application of large electric fields without drawing large currents. The electric field across the quantum wells is controlled by applying a reverse bias to the p–i–n diode, which then also operates as a photodetector with a gain close to unity. The responsivity of such a p–i–n diode operating as a

Figure 12.8. Illustration of the potential seen by an electron–hole pair in a quantum well.[13] The left side shows the situation without an electric field applied, and the right side with an electric field applied. The top diagrams are for an electric field perpendicular to the plane of the layer, and the bottom diagrams are for an electric field parallel to the plane of the layer.

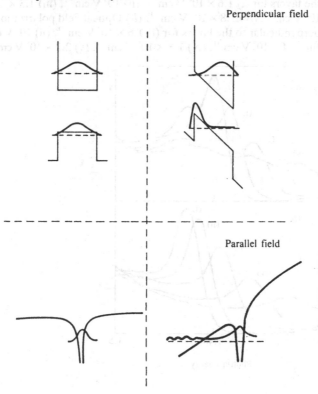

Perpendicular field

Parallel field

photodetector, as a function of photon energy, is shown in Figure 12.10; the curves reproduce the absorption curves that are measured under the same conditions.

Several specific applications utilizing these effects have been proposed.

### High-speed modulation of a light beam

These devices can be used for high-speed modulation of the transmission of a light beam with photon energy just less than the bandgap of the material. Positive results have been made with light transmission both perpendicular and parallel to the layers. Modulation in the 100 ps range has been demonstrated, but it is believed that ultimately response times in the subpicosecond range should be achievable. The devices show promise for optical integrated devices.

Figure 12.9. Exciton absorption in a waveguide containing two 94 Å GaAs quantum wells, as a function of the electric field applied perpendicular to the layers.[13] (a) Optical field polarization parallel to the layers for (i) $1.6 \times 10^4$ V cm$^{-1}$, (ii) $10^5$ V cm$^{-1}$, (iii) $1.3 \times 10^5$ V cm$^{-1}$, (iv) $1.8 \times 10^5$ V cm$^{-1}$. (b) Optical field polarization perpendicular to the layers for (i) $1.6 \times 10^4$ V cm$^{-1}$, (ii) $10^5$ V cm$^{-1}$, (iii) $1.4 \times 10^5$ V cm$^{-1}$, (iv) $1.8 \times 10^5$ V cm$^{-1}$, (v) $2.2 \times 10^5$ V cm$^{-1}$.

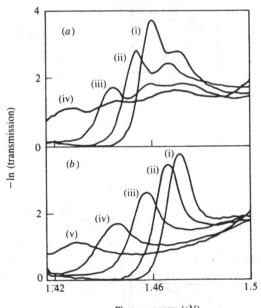

Photon energy (eV)

### Wavelength-selective voltage-tunable photodetector

The results shown in Figure 12.10 demonstrate that a quantum well device can be used to measure absorption by measuring the photocurrent of the device.[14,15] Typical use would be as follows. Light with photon energy less than that of the heavy hole exciton at zero applied electric field is used to illuminate the p–i–n diode, and an electric field is applied to maximize the measured photocurrent by shifting the exciton absorption to the energy of the incident light. Changes in the photon energy of the incident light can be followed by applications of an electric field just sufficient to maximize the current again. The device therefore provides a means of following the frequency of the light electrically.

A p–i–n diode with 50 GaAs quantum wells, each 95 Å wide in the i-region, has been tested to show a wavelength precision of 0.03 Å and a minimum detectable photopulse width of about 200 ps.[14] Similar results have been reported for a superlattice consisting of 100 Å GaAs wells and 50 Å $Al_{0.25}Ga_{0.75}As$ barriers surrounded by thin undoped graded superlattice buffer layers.[15]

Figure 12.10. Responsivity spectra of a p–i–n quantum well diode including layers like those of Figure 12.9 in the i-layer, as a function of applied reverse bias.[13]

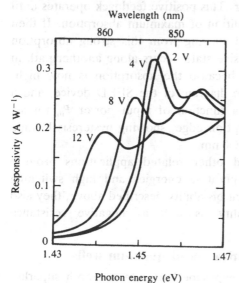

### Self electrooptic effect devices

Self electrooptic effect devices (SEEDs) are based on quantum well p–i–n structures as described above and make use of the ability of these devices to behave as a closely coupled photodetector and modulator with a direct relationship between optical and electronic response.[13] The current passed by a SEED can be affected both by (a) the current vs voltage dependence of the device as an element in a general circuit, and (b) the response of the device to a particular voltage, illuminating light flux $P_{in}$, and light wavelength combination. Since these two requirements must be satisfied at the same time, there arises the possibility of feedback, either positive or negative, between electronic and optical responses. They may be used in optically bistable devices, self-linearized modulators and optical level shifters.[16]

As an example of one type of application, consider the low switching energy, optically bistable gate. The SEED is illuminated by a light beam of variable intensity with wavelength close to the peak of the exciton absorption at zero applied electric field, and a bias is applied to shift the absorption to longer wavelengths. For low light intensities the transmission is relatively high; as the intensity increases, however, the photocurrent increases, which, in turn, induces a voltage drop in the load resistor of the circuit and reduces the electric field across the SEED. As a consequence the absorption increases, as does the photocurrent for the same light intensity, producing an additional voltage drop in the load resistor. This positive feedback operates until the SEED switches to the condition of maximum absorption. If then the light intensity is decreased starting from this strong absorption state, switching to the transmissive state occurs along another path in the variation of $P_{out}$ with $P_{in}$ because the absorption is now high. Figure 12.11 shows a schematic diagram of the SEED device, and a plot of output power $P_{out}$ as a function of input power $P_{in}$ for six wavelengths close to the absorption edge, showing hysteresis leading to bistability between 850 and 860 nm.

SEED devices for this and other related applications promise performance with very low operating energies and high switching speeds. In addition to the optical bistability described above, they also can exhibit electrical bistability as well as negative resistance oscillations.

## 12.6    Internal electric polarization of quantum wells

If a multilayer graded-gap superlattice (a sawtooth superlattice) is illuminated, internal polarization within quantum wells can give rise to a photovoltage detectable in an external circuit.[1] Figure 12.12 shows the energy band diagram for such a sawtooth superlattice, which has been assumed to be p-type and to have a negligible valence-band offset; the layer thicknesses are typically a few hundred angstroms, and the superlattice is bounded by two highly doped p$^+$ contacts.

If electron–hole pairs are excited by a short light pulse, the grading of the wells results in the electrons experiencing a higher electric field than the holes. Under the effects of this electric field and because of their much higher mobility, the electrons separate from the holes and reach the low-gap side of the well in about $10^{-13}$ s. This charge separation causes an electrical polarization in the sawtooth structure that manifests itself as a measurable photovoltage across the contacts. This photovoltage decays in time due both to dielectric relaxation, and to hole drift under the internal polarization-induced field. Thus the current carried in this photodetector is a displacement current, which by continuity is equal to the conduction current in the external circuit.

This effect has been demonstrated in a graded-gap p-type superlattice grown by MBE.[17] Ten graded periods were grown with a period of about 500 Å. The layers were graded from GaAs to $Al_{0.2}Ga_{0.8}As$. For illumination by 4 ps pulses of 6400 Å light, the decay time for the

Figure 12.11. Variation of optical output power $P_{out}$ as a function of optical input power $P_{in}$ for a SEED operating as an optical gate, for six wavelengths near the exciton peak.[13]

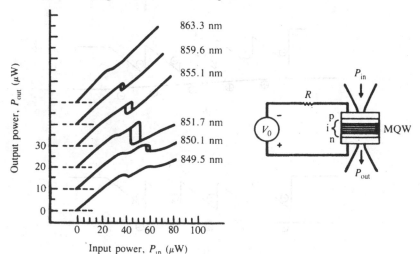

induced photovoltage was about 200 ps. Similar graded-gap superlattice structures have been demonstrated to show photorefractive and non-linear absorption properties.[18]

A somewhat related effect is the observation of photoconductivity gains greater than unity in $InP/In_{0.47}Ga_{0.53}As$ forward-biased multi-quantum well p–n junctions.[19] The interesting characteristic of these devices is that the photoconductivity gain arises not from a modulation of the conductivity, but from a modulation of the electric field. The electric field in the wells is partially screened by the pile-up of photoexcited carriers, leading at constant applied bias voltage to an increase in the field in the barrier layers. In structures with injecting contacts, this mechanism leads to enhanced injection and hence gains exceeding unity. A gain of 27 has been measured for an applied voltage of 3.1 V.

Figure 12.12. Formation and decay of the electrical polarization in a sawtooth superlattice.[1] (*a*) Photoexcitation of electron–hole pairs by a very short light pulse. (*b*) An electrical polarization is set up that manifests itself as a photovoltage on the contacts. (*c*) Dielectric relaxation of the holes restores a flat valence band.

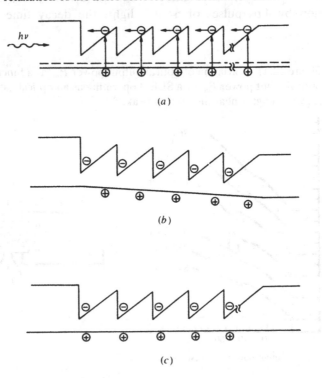

## 12.7 Tunable band-edge discontinuities

In many applications of quantum well superlattices, the magnitudes of the bandgap discontinuities in a heterojunction structure play a critical role. In all of the cases described thus far in this chapter, the magnitude of the bandgap discontinuity is given by the choice of the materials. One step beyond this is the actual designing of this magnitude to meet specific needs. In this section we consider two methods that have been proposed for achieving this degree of 'bandgap engineering.'

### Doping interface dipoles

Very thin sheets (down to monolayers) of ionized donors and acceptors are incorporated by MBE deposition within a few tens of angstroms from the heterojunction interface.[1] Figure 12.13 shows possible effects of such a doping interface dipole (DID) on the conduction band edge in an abrupt heterojunction, assuming that band bending can be neglected.

Depending on the polarity of a sheet of ionized donors and ionized acceptors with identical concentrations introduced during the growth, the band discontinuity may either be increased, as in the top of Figure 12.13, or decreased as in the bottom of Figure 12.13. The band spikes

Figure 12.13. DIDs used to alter the band discontinuities in a heterojunction.[1] (a) Dipole increases the magnitude of the discontinuity. (b) Dipole decreases the magnitude of the discontinuity.

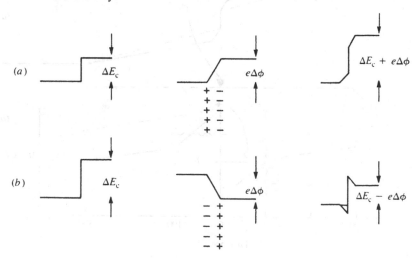

introduced in the bottom of Figure 12.13 may be made totally
transparent by keeping the DID a few atomic layers thick. The doping
density $N$ is between $10^{17}$ and $10^{19}$ cm$^{-3}$ and the sheet thickness $t$ is
kept to less than 100 Å so that the sheets are depleted of carriers.
Acting like a capacitor, the DID shifts the relative positions of the
band edges in the two semiconductors outside the dipole region to
produce abrupt potential variations across the heterojunction interface, without changing the electric field outside the DID. The
electrostatic potential of the 'doping interface dipole' is added to or
subtracted from the dipole potential of the band discontinuity.
Electrons crossing the discontinuity experience a new discontinuity
$\Delta E_c \pm q\Delta\phi$ where $\Delta\phi$ is the potential of the double layer.

To test the effectiveness of these dipole layers in lowering the
barrier height, heterojunction AlGaAs/GaAs p–i–n diodes, one with
a dipole layer and the other without such a layer, were prepared by
MBE on $p$-type (100)GaAs substrates.[1] The one with the dipole
consists of four GaAs layers with the following doping and thicknesses

Figure 12.14. (*a*) Band diagram of the p–i–n junction not drawn to
scale, with solid lines representing the junction with the dipole, and
dashed lines the junction without the dipole. (*b*) Scale band diagram
near the junction interface with and without the dipole.[1]

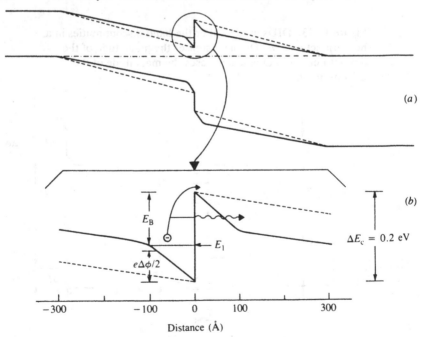

respectively: (1) Be-doped $p^+ > 10^{18}\,\mathrm{cm}^{-3}$ (5000 Å), (2) undoped (5000 Å), (3) Be-doped $p^+ = 5 \times 10^{17}\,\mathrm{cm}^{-3}$ (100 Å) – the negatively charged sheet of the dipole, and (4) undoped (100 Å); these are followed by four $\mathrm{Al}_{0.26}\mathrm{Ga}_{0.74}\mathrm{As}$ layers: (1) undoped (100 Å), (2) Si-doped $n^+ = 5 \times 10^{17}\,\mathrm{cm}^{-3}$ (100 Å) – forming the positively charged sheet of the dipole, (3) undoped (5000 Å), and (4) Si-doped $n^+ \geq 10^{18}\,\mathrm{cm}^{-3}$ (5000 Å). Undoped layers had a carrier density of $\leq 10^{14}\,\mathrm{cm}^{-3}$. The second p–i–n junction without the dipole is identical to the first but without this dipole.

Figure 12.14 shows the band diagram of the p–i–n diodes with and without the dipole, with solid lines for the case with the dipole and dashed lines for the case without. Figure 12.14(*b*) is drawn to scale. The potential of the dipole $\Delta\phi = 0.14\,\mathrm{V}$. The barrier height with the dipole is about one-half of its value without the dipole. It would be expected that major differences should be detectable in the photo-collection efficiency of the two junctions.

Figure 12.15 shows the external quantum efficiency of the two p–i–n junctions deduced from a measurement of the short-circuit current for light incident on the AlGaAs side of the junction. The figure shows that the quantum efficiency $\eta$ is very low (of the order of 2%) for

Figure 12.15. Comparison of the spectral response of external quantum efficiency of the junctions with and without a dipole for zero bias voltage.[1]

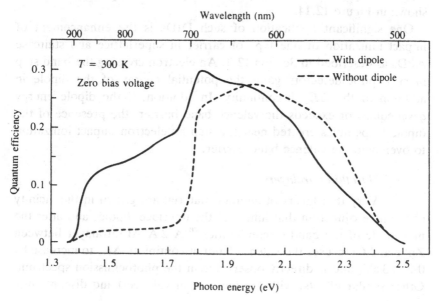

photon energies less than 1.75 eV for the junction without a dipole. This energy corresponds to the bandgap of the $Al_{0.26}Ga_{0.74}As$ layer. For photon energies between 1.40 and 1.75 eV, most of the photoexcited electrons reach the junction interface since incident photons are absorbed partly in the GaAs electric field region and partly in the $p^+$ GaAs layer within a diffusion length from the depletion layer. At the interface these electrons must overcome a 0.2 eV potential barrier in order to be able to contribute to the measured photocurrent. Since this barrier height is much larger than $kT$, the quantum efficiency for these photons is very small. For photon energies larger than 1.75 eV, the light is absorbed primarily in the AlGaAs; since these photoexcited electrons do not have to overcome the barrier at the interface, the quantum efficiency increases. For photon energies larger than 2.0 eV, recombination in the $n^+$ AlGaAs layer and at the surface begin to become important, and the quantum efficiency starts to decrease.

Figure 12.15 also shows that for the junction with a dipole, the quantum efficiency for photon energies greater than 1.75 eV does not differ greatly from those for the junction without a dipole, but for photon energies less than 1.75 eV, where the quantum efficiency in the junction without a dipole was interface barrier limited, the quantum efficiency in the junction with a dipole is almost an order of magnitude larger. Reduction of the interface barrier by about 87 meV by the dipole results in enhanced thermionic emission over the barrier, as well as the possibility of tunneling through the thin triangular barrier shown in Figure 12.14.

One significant application of such DIDs is the enhancement of impact ionization of one type of carrier in superlattice and staircase APDs as described in Section 12.3. An electron crossing the band step in such photodetectors gains the potential energy of the dipole in addition to the $\Delta E_c$ discontinuity. In addition, if the dipole energy $q\Delta\phi$ equals or exceeds the valence band barrier, the presence of the dipole helps holes created near the step by electron impact ionization to overcome the valence band barrier.

### Ultrathin intralayers

Very thin layers of another material are grown in the vicinity of the heterojunction that influence the interface dipoles and alter the magnitude of the band discontinuities.[20] A 2 Å Al intralayer between ZnSe and Ge causes the valence band discontinuity $\Delta E_v$ to increase by 0.2–0.3 eV, and is directly observable in the photoemission spectrum. Other similar effects, with the increase in valence band discontinuity given in parentheses, include Al in CdS–Ge (0.1–0.2 eV), Al in CdS–Si (0.1–0.3 eV), Cs in $SiO_2$–Si (0.2–0.3 eV), Cs in CdS–Si (0.2 eV), Cs in GaP–Ge (0.3–0.4 eV), and $H_2$ in $SiO_2$–Si (±0.5 eV depending on interface preparation).[21]

## 12.8 Tunneling-related phenomena

The effective mass filtering photoconductor discussed in Section 12.2 is only one example of many in which tunneling plays an important role in the properties of quantum wells and superlattices. Other similar phenomena involving resonant tunneling have been suggested for transistors, solid-state lasers, and infrared photodetectors.[3,22]

The principle of resonant tunneling is illustrated in Figure 12.16 for the case of a double barrier with applied dc voltage. The source of electrons is from the heavily doped contact region to the left. When the discrete level in the region between the barriers corresponds to the same energy as the Fermi energy in the highly conducting contact material to the left, electrons tunnel through the double barrier. If the current vs voltage curve is measured, maxima are observed as tunneling occurs through different discrete levels, giving rise to regions of negative resistance potentially useful in oscillators.

### Sequential resonant tunneling

In superlattices with relatively thick barriers (≥100 Å), localized electron states exist, and it is possible to see sequential resonant tunneling between different wells as shown in Figure 12.17. When changes with applied bias occur such that the energy of the ground state $E_1$ in well $n$ is equal to the energy of the first excited state $E_2$ in well $(n + 1)$, sequential resonant tunneling proceeds from well to well. With further increase in applied bias, the energy of the ground state $E_1$ may become equal to the energy of the second excited state $E_3$; a

Figure 12.16. Resonant tunneling through a double barrier heterostructure.[3]

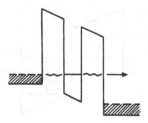

measurement of current through the device as a function of applied voltage therefore shows a series of maxima corresponding to the resonance steps. In the latter case, if the transition from state $E_3$ to state $E_2$ inside the well is radiative, then the potentiality for a tunneling laser exists.

Measurements demonstrating this effect are shown in Figure 12.18.[22] A $p^+$–i–$n^+$ diode was used with the i-layer consisting of a 35 period, undoped ($n^- < 10^{14} \, \text{cm}^{-3}$)$Al_{0.48}In_{0.52}As$ (139 Å), $Ga_{0.47}In_{0.53}As$ (139 Å) superlattice grown by MBE, which was completely depleted at zero applied bias. The electron photocurrent, generated by photoexcitation of the $p^+$ contact layer, was measured as a function of applied reverse bias for different temperatures. The structure shown in Figure 12.18 corresponds to resonant transitions from ground state to first excited state, and from ground state to second excited state, in sequential wells. Hole photocurrent, generated by photoexcitation of the $n^+$ contact layer showed no such structure since resonant tunneling of holes in these superlattices is too weak to be observed. The rising

Figure 12.17. (*a*) Sequential resonant tunneling through a superlattice with transitions inside wells being non-radiative. (*b*) Sequential resonant tunneling involves radiative transitions inside wells, leading to a tunneling laser.[3]

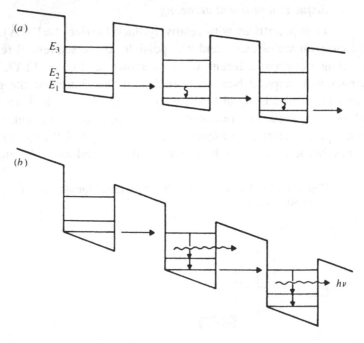

background photocurrent seen in Figure 12.18 is caused by hot electrons transported by thermionic emission over the barriers rather than by tunneling.

Another way of observing the consequences of sequential resonant tunneling is to measure the spectral response of such a $p^+-i-n^+$ junction. Such measurements have been reported for a 50 period $Al_{0.48}In_{0.52}As$, $Ga_{0.47}In_{0.53}As$ i-layer superlattice with 103 Å thick layers,[22] and are shown in Figure 12.19. Three plateaus corresponding to the step-like density of states are seen, corresponding to transitions between the $n = 1$, 2, and 3 valence and conduction subbands, as well as well-defined $n = 1$ and $n = 2$ exciton peaks (even at room temperature) corresponding to transitions involving light and heavy roles.

### Far infrared laser

An example of a suggested application for subsequent resonant tunneling is a far infrared laser, pictured in Figure 12.20.[22] Two variations are shown in the figure. Figure 12.20($a$) shows the situation of photon-assisted tunneling, in which an electron in the ground state of the $n$th well tunnels to an excited state in the $(n + 1)$th well with the simultaneous emission of a photon. Figure 12.20($b$) shows the case where the photon is emitted between states in a particular well after the tunneling has been completed. It appears that the population inversion required for laser activity is more easily achieved with ($a$) than with ($b$).

Figure 12.18. Photocurrent as a function of applied voltage for a $p^+-i-n^+$ junction with a superlattice i-layer as described in the text.[22]

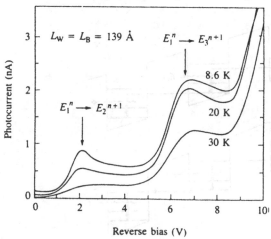

Figure 12.19. Spectral response at room temperature of a $p^+-i-n^+$ photodiode with a superlattice i-layer. (HH = heavy holes; LH = light holes)[22]

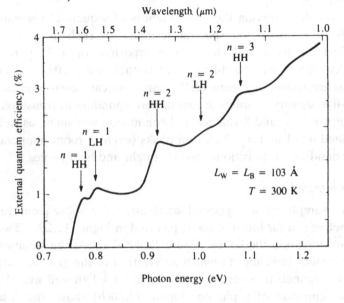

Figure 12.20. Two possibilities proposed for a subsequent resonant tunneling far infrared laser.[22] (*a*) Photon is emitted during an interwell photon-assisted tunneling. (*b*) Photon is emitted during an intrawell transition between excited states after tunneling between wells.

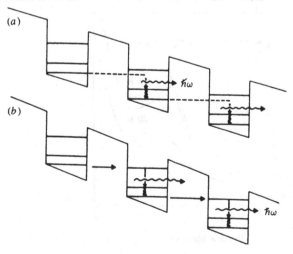

### Wavelength-selective infrared photodetector

The energy band diagram for a wavelength-selective infrared photodetector based on sequential resonant tunneling in a superlattice is given in Figure 12.21. A materials description would call for n-type quantum wells with thicknesses in the 100 Å range, undoped barriers with thickness in the 20–50 Å range, and $n^+$ contacts on both sides. A bias voltage is selected so that the second excited level in the $n$th well has the same energy as the third excited level in the $(n + 1)$th well. Photons incident on the device, with a component of the electric field perpendicular to the plane of the well, will be absorbed only if the photon energy is very close to $(E_2 - E_1)$. Typical half maximum linewidths at room temperature are 10 meV, thus providing high wavelength selectivity.

After photoexcitation from the ground state to the first excited state of the $n$th well, the electron can tunnel to the second excited state of the $(n + 1)$th well. After tunneling the electron can relax to the first excited state, whereupon it can again tunnel to the second excited state of the $(n + 2)$th well, or it can relax to the ground state of the $(n + 1)$th well. As a result of these various processes, a photocurrent results whenever the electron relaxes to the ground state in a different well from the one in which it was initially excited. The quantum efficiency of the detector will be optimized by insuring that the barriers are sufficiently thin that the tunneling time from the first to the second excited state between wells is smaller than the scattering time to the ground state. For 140 Å wells in $Al_{0.48}In_{0.52}As/Ga_{0.47}In_{0.53}As$ structures, this criterion calls for a barrier thickness less than 40 Å. Indications have been reported of the involvement of deep imperfection levels in the barrier layers as part of an enhanced tunneling process.[23]

Figure 12.21. An infrared photoconducting detector utilizing sequential resonant tunneling.[22]

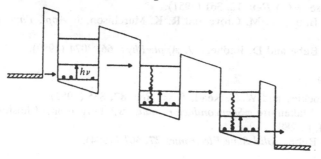

# Notes

## Chapter 1

1. For a detailed exposition of the quantum mechanical calculation involved in optical absorption, see for example Sections 11.3 and 11.4 in R. H. Bube, *Electronic properties of crystalline solids,* Academic Press, NY (1974).
2. See, for example, the 'delta function model,' G. Lucovsky, *Solid State Comm.* **3,** 299 (1965); the 'quantum defect model,' H. B. Bebb and R. A. Chapman, *J. Phys. Chem. Solids* **28,** 2087 (1967); 'non-parabolic bands,' H. G. Grimmeiss and L-A. Ledebo, *J. Phys. C: Solid St. Phys.* **8,** 2615 (1975); 'modified spherical square-well,' J. W. Walker and C. T. Sah, *Phys. Rev.* **B8,** 5597 (1973).
3. See, for example, R. H. Tredgold, *Space charge conduction in solids,* Elsevier, NY (1966); M. A. Lampert and P. Mark, *Current injection in solids,* Academic Press, NY (1970).

## Chapter 2

1. R. H. Bube and L. A. Barton, *RCA Rev.* **20,** 564 (1959).
2. H. B. DeVore, *Phys. Rev.* **102,** 86 (1956).
3. A. M. Goodman, *J. Appl. Phys.* **30,** 144 (1959).
4. J. Bornstein and R. H. Bube, *J. Appl. Phys.* **61,** 2676 (1987).

## Chapter 3

1. M. D. Tabak and P. J. Warter, *Phys. Rev.* **148,** 982 (1966).
2. A. Rose, *RCA Rev.* **12,** 361 (1951).
3. R. H. Bube, W. M. Grove and R. K. Murchison, *J. Appl. Phys.* **38,** 3515 (1967).
4. R. H. Bube and D. Redfield, *J. Appl. Phys.* **66,** 3074 (1989).

## Chapter 4

1. W. Shockley and W. T. Read, *Phys. Rev.* **87,** 835 (1952).
2. J. S. Blakemore, *Semiconductor statistics,* Pergamon, Elmsford, NY, (1962), p. 280.
3. R. H. Bube, *Solid-State Electronics* **27,** 467 (1984).

*Chapter 5*

1. P. Goercke, *Ann. Telecommun.* **6**, 325 (1951).
2. R. Newman, H. H. Woodbury, and W. W. Tyler, *Phys. Rev.* **102**, 613 (1956).
3. S. O. Hemila and R. H. Bube, *J. Appl. Phys.* **38**, 5258 (1967).
4. R. H. Bube and F. Cardon, *J. Appl. Phys.* **35**, 2712 (1964).
5. R. H. Bube and C-T. Ho, *J. Appl. Phys.* **37**, 4132 (1966).
6. R. H. Bube, *J. Phys. Chem. Solids* **1**, 234 (1957).
7. R. H. Bube, *Solid State Electronics* **27**, 467 (1984).
8. I. Broser, in *Physics and chemistry of II–VI compounds,* M. Aven and J. S. Prener, eds., North-Holland Publ., Amsterdam, 533 (1967).
9. G. B. Stringfellow and R. H. Bube, *Phys. Rev.* **171**, 903 (1968).
10. R. H. Bube, *J. Appl. Phys.* **35**, 586 (1964).
11. F. F. Morehead, *J. Phys. Chem. Solids* **24**, 37 (1963).
12. R. H. Bube and A. B. Dreeben, *Phys. Rev.* **115**, 1578 (1959).
13. G. A. Dussel and R. H. Bube, *J. Appl. Phys.* **37**, 13 (1966).
14. A. F. Sklensky and R. H. Bube, *Phys. Rev.* **6**, 1328 (1972).


*Chapter 6*

1. R. H. Bube and F. Cardon, *J. Appl. Phys.* **35**, 2712 (1964).
2. H. Gummel and M. Lax, *Phys. Rev.* **97**, 1469 (1955).
3. M. Lax, *Phys. Rev.* **119**, 1502 (1960).
4. C. H. Henry and D. V. Lang, *Phys. Rev.* **15**, 989 (1975).
5. E. Burstein, *Photoconductivity conference,* Wiley, NY (1956), p. 353.
6. J. Saura and R. H. Bube, *J. Appl. Phys.* **36**, 3660 (1965).


*Chapter 7*

1. V. A. Johnson and K. Lark-Horovitz, *Phys. Rev.* **82**, 977 (1951).
2. S. S. Devlin in *Physics and chemistry of II–VI compounds,* M. Aven and J. Prener, eds., Wiley, NY, (1967), p. 561.
3. N. M. Amer and W. B. Jackson, in *Semiconductors and semimetals: Hydrogenated amorphous silicon* **21B**, J. L. Pankove, ed., Academic Press, NY (1984), p. 83.
4. M. Vanecek, J. Kocka, J. Stuchlik and A. Triska, *Solid State Comm.* **39**, 1199 (1981).
5. A. L. Fahrenbruch and R. H. Bube, *Fundamentals of solar cells,* Academic Press, NY (1983), pp. 90–102.
6. T. S. Moss, *Proc. Phys. Soc.* **66B**, 993 (1953); *Physica* **20**, 989 (1954).
7. A. M. Goodman, *J. Appl. Phys.* **32**, 2550 (1961).
8. A. R. Moore, in *Semiconductors and semimetals: Hydrogenated amorphous silicon* **21C**, J. L. Pankove, ed., Academic Press, NY (1984), p. 239.
9. F. Berz and H. K. Kuiken, *Solid-State Electron.* **19**, 437 (1976).
10. See Ref. 5, p. 81.
11. J. E. Mahan, T. W. Ekstedt, R. I. Frank, and R. Kaplow, *IEEE Trans. Electron Devices* **ED-26**, 733 (1979).

12. F. P. Heiman, *IEEE Trans. Electron Devices* **ED-14,** 781 (1967).
13. H. E. MacDonald and R. H. Bube, *Rev. Sci. Instr.* **33,** 721 (1962).
14. R. H. Bube and H. E. MacDonald, *Phys. Rev.* **121,** 473 (1961); **128,** 2062 (1962); **128,** 2071 (1962).
15. R. H. Bube, H. E. MacDonald and J. Blanc, *J. Phys. Chem. Solids* **22,** 173 (1961).
16. R. H. Bube, *Scientia, Sixieme Serie,* January (1964).
17. A. L. Lin, E. Omelianovski, and R. H. Bube, *J. Appl. Phys.* **47,** 1852 (1976).
18. R. H. Bube in *Physical chemistry.* Vol. X. *Solid state,* H. Eyring, D. Henderson, and W. Jost, eds. Academic Press, NY (1970), p. 515.
19. H. B. Kwok and R. H. Bube, *J. Appl. Phys.* **44,** 138 (1973).
20. S. S. Chiao, B. L. Mattes and R. H. Bube, *J. Appl. Phys.* **49,** 261 (1978).
21. G. Lucovsky, *Solid State Commun.* **3,** 299 (1965).
22. G. B. Stringfellow and R. H. Bube, *Phys. Rev.* **171,** 903 (1968).
23. G. B. Stringfellow and R. H. Bube, *J. Appl. Phys.* **39,** 3657 (1968).
24. D. Ritter, E. Zeldov and K. Weiser, *Appl. Phys. Lett.* **49,** 791 (1986).

*Chapter 8*

1. There may be situations, indeed, in which $\tau_0$ is a function of time, as when there is a distribution of activation energies, capture coefficients or some other time-dependent critical factor. If $\tau_0$ can be represented as a power-law function of time $\tau_0 \propto t^\alpha$, then Eq. (8.1) is converted into a 'stretched exponential' with $n \propto \exp[-(t/\tau_{00})^{1-\alpha}]$. Several applications of such stretched exponentials have occurred in recent analyses of phenomena in amorphous semiconductors and solid-solution semiconductor systems, particularly phenomena involving structural reorganization of imperfections as part of the decay process. See, e.g., R. H. Bube, L. Echeverria and D. Redfield, *Appl. Phys. Lett.* **57,** 79 (1990).
2. P. Besomi and B. Wessels, *J. Appl. Phys.* **51,** 4305 (1980).
3. R. H. Bube, W. M. Grove and R. K. Murchison, *J. Appl. Phys.* **38,** 3515 (1967).
4. R. H. Bube, *J. Appl. Phys.* **32,** 1621 (1961).
5. S. W. S. McKeever, *Thermoluminescence of solids,* Cambridge University Press, Cambridge (1985).
6. The reader will hopefully be indulgent enough to allow me to slip in these figures from my Ph.D. thesis. R. H. Bube, *Phys. Rev.,* **80,** 655 (1950).
7. R. H. Bube, *Phys. Rev.* **81,** 633 (1951).
8. R. H. Bube, G. A. Dussel, C-T. Ho, and L. D. Miller, *J. Appl. Phys.* **37,** 21 (1966).
9. R. H. Bube, in *Physical chemistry.* Volume X, *Solid state,* H. Eyring, D. Henderson and W. Jost, eds., Academic Press, NY (1970), p. 515.
10. R. H. Bube and H. E. MacDonald, *Phys. Rev.* **128,** 2071 (1962).
11. E. Tscholl, *Philips Res. Rep. Supplements,* No. 6 (1968).
12. H. B. Im, H. E. Matthews and R. H. Bube, *J. Appl. Phys.* **41,** 2581 (1970).
13. C. Hurtes, M. Boulou, A. Mitonneau, and D. Bois, *Appl. Phys. Lett.* **32,** 821 (1978).
14. D. V. Lang, *J. Appl. Phys.* **45,** 3023 (1974).

15. G. L. Miller, D. V. Lang, and L. C. Kimerling, *Ann. Rev. Mater. Sci.* **7**, 377 (1977).
16. L. C. Kimerling, *Int. conf. on radiation effects in semiconductors, Dubrovnik,* Institute of Physics, Bristol (1976).
17. D. V. Lang, R. A. Logan, and L. C. Kimerling, *Proc. XIII int. conf. on the physics of semiconductors, Rome,* North-Holland, Amsterdam (1976), p. 615.
18. G. L. Miller, *IEEE Trans. Electron. Devices* **ED-19**, 1103 (1972).
19. N. M. Johnson, D. J. Bartelink, R. B. Gold, and J. F. Gibbons, *J. Appl. Phys.* **50**, 4828 (1979).
20. M. Li and C. Sah, *IEEE Trans. Electron. Devices* **ED-29**, 306 (1982).
21. J. J. Shiau, A. L. Fahrenbruch, and R. H. Bube, *J. Appl. Phys.* **61**, 1340 (1987).

*Chapter 9*

1. W. Huber, A. L. Fahrenbruch, C. Fortmann and R. H. Bube, *J. Appl. Phys.* **54**, 4038 (1983).
2. H. C. Card and E. S. Yang, *IEEE Trans. Electron. Devices* **ED-24**, 397 (1977).
3. C. H. Seager, *J. Appl. Phys.* **52**, 3960 (1981).
4. J. Y. Seto, *J. Appl. Phys.* **46**, 5247 (1975).
5. S. M. Sze, *Physics of semiconductor devices,* Wiley, NY (1969), p. 382.
6. T. P. Thorpe, A. L. Fahrenbruch and R. H. Bube, *J. Appl. Phys.* **60**, 3622 (1986).
7. G. E. Pike and C. H. Seager, *J. Appl. Phys.* **50**, 3414 (1979).
8. J. Werner, W. Jantsch, K. H. Froehner and H. J. Queisser, in *Grain boundaries in semiconductors,* H. J. Leamy, G. E. Pike, and C. H. Seager, eds, North-Holland, NY (1982), p. 101.
9. C. Wu and R. H. Bube, *J. Appl. Phys.* **45**, 648 (1974).
10. Y. Y. Ma and R. H. Bube, *J. Electrochem. Soc.* **124**, 1430 (1977).
11. H. Shear, E. A. Hilton and R. H. Bube, *J. Electrochem. Soc.* **112**, 997 (1965).
12. A. L. Robinson and R. H. Bube, *J. Electrochem. Soc.* **112**, 1001 (1965).
13. R. H. Bube, *J. Electrochem. Soc.* **113**, 793 (1966).
14. S. Espevik, C. Wu and R. H. Bube, *J. Appl. Phys.* **42**, 3513 (1971).
15. R. H. Bube, *Ann. Rev. of Mat. Sci.* **5**, 201 (1975).
16. G. H. Blount, R. H. Bube, and A. L. Robinson, *J. Appl. Phys.* **41**, 2190 (1970).
17. R. L. Petritz, F. L. Lummis, H. E. Sorrows, and J. F. Woods, *Semiconductor surface physics,* University of Pennsylvania Press, Philadelphia (1957), p. 229.

*Chapter 10*

1. P. G. LeComber and W. E. Spear, *Phys. Rev. Lett.* **25**, 509 (1970). See also W. E. Spear and P. G. LeComber, *J. Non-Cryst. Solids* **8–10**, 727 (1972); *Solid State Commun.* **17**, 1193 (1975); *Phil. Mag.* **33**, 935 (1976).
2. J. I. Pankove, ed., *Semiconductors and semimetals.* Vol. 21. *Hydrogenated amorphous silicon.* Parts A, B, C, and D. Academic Press, NY (1984).

3. See, e.g., N. F. Mott and E. A. Davis, *Electronic properties in non-crystalline materials,* Oxford University Press, London (1971).
4. N. M. Amer and W. B. Jackson, Chapter 3 in *Semiconductors and semimetals* Vol. 21. *Hydrogenated amorphous silicon.* Part B. *Optical properties,* J. I. Pankove, ed., Academic Press, NY (1984).
5. R. H. Bube, *RCA Rev.* **36,** 467 (1975).
6. R. H. Bube, J. E. Mahan, R. T.-S. Shiah and H. A. Vander Plas, *Appl. Phys. Lett.* **25,** 419 (1974).
7. T. C. Arnouldussen, R. H. Bube, E. A. Fagen and S. Holmberg, *J. Appl. Phys.* **43,** 1798 (1972); *J. Non-Cryst. Solids* **8–10,** 933 (1972).
8. J. G. Simmons and G. W. Taylor, *J. Non-Cryst. Solids* **8–10,** 947 (1972); *J. Phys. C: Solid State Phys.,* **6,** 3706 (1973).
9. G. W. Taylor and J. G. Simmons, *J. Phys. C: Solid State Phys.,* **7,** 3067 (1974).
10. T. C. Arnoldussen, C. A. Menezes, Y. Nakagawa, and R. H. Bube, *Phys. Rev.* **9,** 8, 3377 (1974).
11. R. T.-S. Shiah and R. H. Bube, *J. Appl. Phys.* **47,** 2005 (1976).
12. J. E. Mahan and R. H. Bube, *J. Non-Cryst. Solids* **24,** 29 (1977).
13. H. A. Vander Plas and R. H. Bube, *J. Non-Cryst. Solids* **24,** 377 (1977).
14. D. L. Staebler and C. R. Wronski, *J. Appl. Phys.* **51,** 3262 (1980).
15. R. H. Bube and D. Redfield, *J. Appl. Phys.* **66,** 3074 (1989).
16. P. G. LeComber and W. E. Spear, in *Amorphous semiconductors,* Vol. 36 of *Topics in applied physics,* M. H. Brodsky, ed., Springer, NY (1986), p. 251.
17. R. A. Street, *J. Non-Cryst. Solids* **77&78,** 1 (1985).
18. M. Stutzmann, W. Jackson, and C. Tsai, *Phys. Rev.* **B32,** 23 (1985).
19. H. Curtins, M. Favre, L. Ziegler, N. Wyrsch, and A. V. Shah, in *Proceedings of the materials research society conference, Reno, NV,* 5–9 April, 1988, Materials Research Society, Pittsburgh PA.

*Chapter 11*

1. See, e.g., A. L. Fahrenbruch and R. H. Bube, *Fundamentals of solar cells: Photovoltaic solar energy conversion,* Academic Press, NY (1983).
2. R. H. Bube and A. L. Fahrenbruch, *Photovoltaic effect* in *Advances in electronics and electron physics* Vol. **56,** C. Marton, ed., Academic Press, NY (1981), p. 163.
3. R. H. Bube, 'Solar cells' in *Handbook on semiconductors,* T. S. Moss, ed., Vol. 4, *Device physics,* C. Hilsum, ed., North-Holland, NY (1981), p. 691.
4. R. H. Bube, 'Materials for photovoltaics' in *Annual Review of Materials Science,* **20,** 19 (1990).
5. R. L. Anderson, *Solid State Electron.* **5,** 341 (1962).
6. H. F. MacMillan, H. C. Hamaker, N. R. Kaminar, M. S. Kuryla, M. Ladle Ristow, *et al.* 1988, *20th IEEE photovoltaic specialist conf., Las Vegas,* IEEE Publishing, NY, p. 462.
7. D. E. Carlson, 'Solar cells' in *Semiconductors and semimetals.* Vol. 21. *Hydrogenated amorphous silicon,* Part D. *Device applications,* J. I. Pankove, ed., Academic Press, NY (1984), p. 7.

8. P. V. Meyers, *Photovoltaic advanced research and development*. 1989. *9th Annu. Rev. Meet.* SERI/CP-213-3495, Lakewood, CO, p. 19.
9. K. W. Mitchell, A. L. Fahrenbruch, and R. H. Bube, *J. Appl. Phys.* **48**, 4365 (1977).
10. A. R. Riben and D. L. Feucht, *Solid State Electron.* **9**, 1055 (1966); *Int. J. Electron.* **20**, 583 (1966).
11. F. A. Padovani and R. Stratton, *Solid State Electron.* **9**, 695 (1966).
12. M. G. Peters, A. L. Fahrenbruch and R. H. Bube, *J. Appl. Phys.* **64**, 3106 (1988).
13. J. A. Aranovich, D. Golmayo, A. L. Fahrenbruch and R. H. Bube, *J. Appl. Phys.* **51**, 4260 (1980).
14. C. Eberspacher, A. L. Fahrenbruch and R. H. Bube, *17th IEEE photovoltaic specialists conference* IEEE, NY (1984), p. 459.
15. J. M. Gee and G. F. Virshup, *20th IEEE photovoltaic specialists conf.*, *Las Vegas* (1988), IEEE Publishing, NY, p. 754.
16. B. J. Stanbery, J. E. Avery, R. M. Burgess, W. S. Chen and W. E. Devaney, *et al.* *19th IEEE photovoltaic specialists conf.*, *New Orleans* (1987), IEEE Publishing, NY, p. 280.
17. N. P. Kim, R. M. Burgess, B. J. Stanbery, R. A. Mickelsen, J. E. Avery, *et al.*, *20th IEEE photovoltaic specialists conf.*, *Las Vegas* (1988), IEEE Publishing, NY, p. 457.
18. C. R. Lewis, H. F. MacMillan, B. C. Chung, G. F. Virshup, D. D. Liu, *et al.* *Solar Cells* **24**, 171 (1988).
19. G. F. Virshup, B. C. Chung, and J. G. Werthen, *20th IEEE photovoltaic specialists conf.*, *Las Vegas*, IEEE Publishing, NY, (1988), p. 441.
20. H. F. MacMillan, B. C. Chung, L. D. Partain, J. C. Schultz, G. F. Virshup, and J. G. Werthen, *9th Annual Rev. Meet.* SERI/CP-213-3495, Lakewood, CO (1989) p. 53.
21. K. W. Mitchell, C. Eberspacher, J. Ermer, and D. Pier, *20th IEEE photovoltaic specialists conf.*, *Las Vegas*, IEEE Publishing, NY, 1988, p. 1384.
22. S. Guha, *9th Annual Rev. Meet.* SERI/CP-213-3495, Lakewood, CO (1989) p. 9.
23. J. M. Olson, S. R. Kurtz, A. E. Kibbler, and E. Beck, *9th Annual Rev. Meet.* SERI/CP-213-3495, Lakewood, CO (1989), p. 51.
24. W. D. Gill and R. H. Bube, *J. Appl. Phys.* **41**, 3731 (1970).
25. P. F. Lindquist and R. H. Bube, *J. Appl. Phys.* **43**, 2839 (1972).
26. S. Kanev, V. Stoyanov, and M. Lakova, *Dok. Bolg. Akad. Nauk.* **22**, 863 (1969).
27. A. L. Fahrenbruch and R. H. Bube, *J. Appl. Phys.* **45**, 1264 (1974).

*Chapter 12*

1. F. Capasso, in *Heterojunction band discontinuities, physics and device applications*, F. Capasso and G. Margaritondo, eds., North-Holland, Elsevier Science Publishing Co., NY (1987), p. 401.
2. F. Capasso, ed., *Physics of quantum electron devices*, Springer-Verlag, Berlin (1990).
3. F. Capasso, *Science* **235**, 172 (1987).

4. F. Capasso and S. Datta, *Physics Today*, Feb. 1990, p. 2.
5. F. Capasso, K. Mohammed, A. Y. Cho, R. Hull, and A. L. Hutchinson, *Appl. Phys. Lett.* **47**, 420 (1985).
6. F. Capasso, W. T. Tsang, A. L. Hutchinson, and G. F. Williams, *Appl. Phys. Lett.* **40**, 38 (1982).
7. F. Capasso, J. Allam, A. Y. Cho, K. Mohammed, R. J. Malik, A. L. Hutchinson and D. Sivco, *Appl. Phys. Lett.* **48**, 1294 (1986).
8. J. Allam, F. Capasso, K. Alavi and A. Y. Cho, *IEEE Electron Device Lett.* **EDL-81**, 2013 (1989).
9. B. F. Levine, G. Hasnain, C. G. Bethea, and N. Chand, *Appl. Phys. Lett.* **54**, 2704 (1989).
10. B. F. Levine, C. G. Bethea, G. Hasnain, V. O. Shen, E. Pelve, R. R. Abbott, and S. J. Hsieh, *Appl. Phys. Lett.* **56**, 851 (1990).
11. G. Hasnain, B. F. Levine, D. L. Sivco and A. Y. Cho, *Appl. Phys. Lett.* **56**, 770 (1990).
12. G. Hasnain, G. F. Levine, S. Gunapala, and N. Chand, *Appl. Phys. Lett.* **57**, 608 (1990).
13. D. S. Chemla and D. A. B. Miller, in *Heterojunction band discontinuities, physics and device applications,* F. Capasso and G. Margaritondo, eds., North-Holland, Elsevier Science Publishing Co., NY (1987), p. 597.
14. T. H. Wood, C. A. Burrus, A. H. Gnauck, J. M. Wiesenfeld, D. A. B. Miller, D. S. Chemla and T. C. Damen, *Appl. Phys. Lett.* **47**, 190 (1985).
15. A. Larsson, A. Yariv, R. Tell, J. Maserjian and S. T. Eng., *Appl. Phys. Lett.* **49**, 821 (1986).
16. D. A. B. Miller, D. S. Chemla, T. C. Damen, T. H. Wood, C. A. Burrus, Jr., A. C. Gossard, and W. Wiegmann, *IEEE J. Quant. Electron.* **QE-21**, 1462 (1985).
17. F. Capasso, S. Luryi, W. T. Tsang, C. G. Bethea, and B. F. Levine, *Phys. Rev. Lett.* **51**, 2318 (1983).
18. S. E. Ralph, F. Capasso and R. J. Malik, *Appl. Phys. Lett.* **57**, 626 (1990).
19. F. Capasso, G. Ripamonti, W. T. Tsang, A. L. Hutchinson, and S. N. G. Chu, *J. Appl. Phys.* **65**, 388 (1989).
20. G. Margaritondo and P. Perfetti, in *Heterojunction band discontinuities, physics and device applications,* F. Capasso and G. Margaritondo, eds., North-Holland, Elsevier Science Publishing Co., NY (1987), p. 59.
21. P. Perfetti, C. Quaresima, C. Coluzza, C. Fortunato and G. Margaritondo, *Phys. Rev. Lett.* **57**, 2065 (1986).
22. F. Capasso, K. Mohammed, and A. Y. Cho, *IEEE J. of Quant. Electron.* **QE-22**, 1853 (1986).
23. F. Capasso, K. Mohammed, and A. Y. Cho, *Phys. Rev. Lett.* **57**, 2303 (1986).

# Bibliography

F. Abeles, ed., *Optical properties of solids*, North-Holland, Amsterdam (1972).

R. G. Breckenridge, B. R. Russell and E. E. Hahn, eds., *Photoconductivity conference*, John Wiley & Sons, NY (1956).

R. H. Bube, 'Photoconductivity,' in *Methods of experimental physics. Vol. 6. Solid state physics*, K. Lark-Horovitz and V. A. Johnson, eds., Academic Press, NY (1959), p. 335.

R. H. Bube, *Photoconductivity of solids*, John Wiley & Sons, NY (1960); Krieger, Huntington, NY (1978).

R. H. Bube, 'Photoelectronic analysis,' in R. K. Willardson and A. C. Beer, eds., *Semiconductors and semimetals*, Vol. 3, *Optical properties of III–V compounds*, Academic Press, NY (1967), p. 461.

R. H. Bube, 'Photoconductivity of semiconductors,' in H. Eyring, D. Henderson and W. Jost, eds., *Physical chemistry*, Vol. X, *Solid state*, Academic Press, NY (1970), p. 515.

R. H. Bube, *Electronic properties of crystalline solids*, Academic Press, NY (1974).

F. Capasso and G. Margaritondo, eds., *Heterojunction band discontinuities: Physics and applications*, North Holland, Elsevier, NY (1987).

F. Capasso, ed., *Physics of quantum electron devices*, Springer-Verlag, NY (1990).

A. L. Fahrenbruch and R. H. Bube, *Fundamentals of solar cells*, Academic Press, NY (1983).

D. L. Greenaway and G. Harbeke, *Optical properties and band structure of semiconductors*, Pergamon Press, Oxford (1968).

N. V. Joshi, *Photoconductivity: Art, science and technology*, Marcel Dekker, Inc., NY (1990).

S. Larach, ed., *Photoelectronic materials and devices*, Van Nostrand, Princeton, NJ (1965).

H. Levinstein, ed., *Photoconductivity*, Pergamon Press, Oxford (1962).

McKeever, S. W. S., *Thermoluminescence of solids*, Cambridge University Press, Cambridge (1985).

J. Mort and D. M. Pai, eds., *Photoconductivity and related phenomena*, Elsevier, NY (1976).

T. S. Moss, *Photoconductivity in the elements*, Academic Press, NY (1952).

E. M. Pell, *Proceedings of the third international conference on photoconductivity*, Pergamon, Oxford (1971).

A. Rose, *Concepts in photoconductivity and allied problems*, Interscience, John Wiley & Sons, NY (1963).

S. M. Ryvkin, *Photoelectric effects in semiconductors*, Consultants Bureau, NY, (1964).

V. S. Vavilov, *Effects of radiation on semiconductors*, Consultants Bureau, NY (1965).

# Index

Printed in the United States
By Bookmasters